Before the Collapse

2/24 4

Ugo Bardi

Before the Collapse

A Guide to the Other Side of Growth

 Springer

Ugo Bardi
Department of Chemistry
University of Florence
Florence, Italy

ISBN 978-3-030-29037-5   ISBN 978-3-030-29038-2   (eBook)
https://doi.org/10.1007/978-3-030-29038-2

This Springer imprint is published by the registered company Springer Nature Switzerland AG
The registered company address is: Gewerbestrasse 11, 6330 Cham, Switzerland

Dedicated to my two granddaughters, Aurora and Beatrice, who were both born while this book was being written and who will live in the future that we can only dimly perceive today.

Foreword by Susan Kucera

I met Ugo Bardi for the first time when I interviewed him in Florence for my 2014 movie "Breath of Life". More recently, he was featured in my 2018 documentary "Living in the Future's Past" and in 2019 we met again in a medieval castle in the hills near Florence for a new film in production. It was a fitting environment to discuss how our future is mirrored in our past. Discussions with Ugo are always fascinating: you find yourself finding parallels between worlds that you would have thought to be so different as to have no points in common. For example, Ugo has such a breadth of knowledge that he can always tell you about how ancient civilizations, from the Sumerians onward, had so many points in common with our world. In particular, Ugo is interested in a comparison of our situation with that of the age that we call "Late Antiquity" or "early Middle Ages".

In those ancient times, people were facing similar problems to those we face today: how can we maintain the achievements of what we call "civilization" in a condition of decline of our material wealth? According to Ugo and his coworker, the young Italian medievalist Alessia Scopece (whom I also met in that medieval castle in 2019), the early Middle Ages were far from being a "dark age". Rather, they were a period of creative adaptation to a difficult economic situation. People living in the Middle Ages developed flexible and inexpensive solutions to problems that were unsolvable within the old paradigms, for instance, lacking precious metals, they developed cultural methods of exchange that replaced conventional methods. According to Ugo, the holy relics that were such a typical feature of the Middle Ages were in many ways to be seen as "money", something that facilitated commerce and travel in Europe.

Ugo is not just interested in the past: he projects into the future and his studies on the great energy transition tell us whether it will be possible to abandon fossil fuels to build a society entirely based on renewable energy. He told me that, "it is obviously possible because it is unavoidable". The problem is not whether we'll get there or not, but how fast and with how much hard work and sacrifices. But just as the Middle Ages were the unavoidable destiny of the declining Roman Empire, a renewable-based society is the unavoidable destiny of our declining civilization.

In this book, Ugo Bardi distils much of his thoughts and his reflections he developed over the past years. It starts from the past, from a thought of the Roman Philosopher Lucius Annaeus Seneca who was perhaps the first in history to note that decline is always faster than thought—"Ruin is rapid", as Seneca wrote. Out of this simple sentence, Ugo builds up a wide-ranging discussion of how we find ourselves in the current plight, desperately trying to fight against forces that we ourselves set in motion and that we are now unable to control. Climate change is the paradigmatic problem of our civilization, one that may very well bring us to that "Seneca Cliff" that Ugo describes in this book.

However, this is not a pessimistic book, it is not a book about doom and gloom, and Ugo is not here to scare us or to tell us that we have no hope to survive. On the contrary, it is a book that gains strength and breadth from the ancient Stoic philosophy of which Seneca was an adept. Stoics understood that the world always changes, sometimes fast, and sometimes so fast that, from our viewpoint, we see the change as a disaster. But all changes happen because they have to happen, and if we'll see big changes in the future it will be because they are necessary. Indeed, the connecting line that goes through this book is what Ugo calls the "Seneca Strategy"—the realization that change is necessary and that in most cases opposing it simply leads to a faster ruin. So, from ancient Stoics, we may learn the wisdom we need to face our uncertain future.

Hawaii   Susan Kucera

Preface: The Seneca Effect: Why Growth is Slow, But Collapse is Rapid

It would be some consolation for the feebleness of our selves and our works if all things should perish as slowly as they come into being; but as it is, increases are of sluggish growth, but the way to ruin is rapid.
Lucius Anneaus Seneca, Letters to Lucilius, n. 91—translation by
Richard M. Gummere

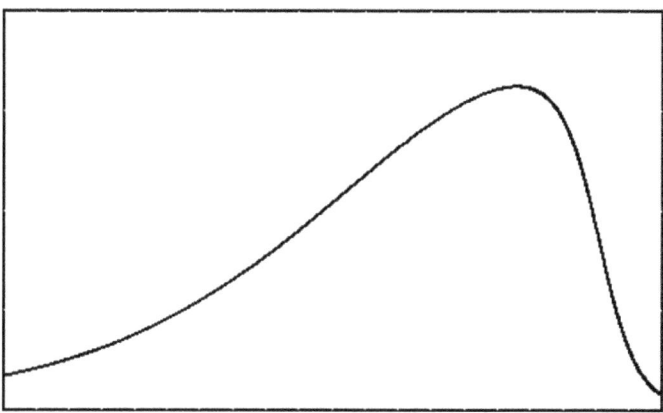

Fig. 1 The "Seneca curve" showing the time evolution of a system. Growth is slow, but the decline is rapid enough that it appears to us as a collapse

Normally, our life is quiet. As ordinary people, we may enjoy moderate prosperity, reasonable happiness, and expected events. But life is also full of surprises and when things start to fail, they tend to fail fast enough for us to use terms such as "collapse", or "ruin", as the Roman philosopher Lucius Annaeus Seneca noted already long ago when he said "increases are of sluggish growth, but the way to ruin is rapid" [1]. And when collapse comes, it often finds us woefully unprepared, that's why we should prepare in advance (Fig. 1).

It may be difficult to define collapses in rigorous terms, but we all can recognize one when we see it. Collapse is a rapid, uncontrolled, unexpected, and ruinous decline of something that had been going well before. It can strike individuals: you may lose your job, or get sick, lose a close friend or a family member. And it can happen very fast, sometimes by chance, sometimes by a mistake: think of the case of Roseanne Barr who, in 2018, saw her career of TV anchor ruined in a day because of a single racist tweet she wrote.

Collapse also affects larger systems. The average lifetime of a commercial company, today, is of the order of 15 years, but small companies tend to come and go much more quickly: it is the "fail fast, fail often" strategy, well known in Silicon Valley and supposed to be a good thing to eliminate the weaklings in the struggle for survival. True, a startup may become a "unicorn", a term coined by venture capitalist Aileen Lee to describe the rare case of a successful startup that reaches a value of over $ 1 billion. For these and even larger companies, demise may be a more difficult and painful affair, sometimes with possibilities of getting back in business as it happened to Evernote, a survivor of the early days of the Web that refuses to go away [2]. But, in most cases when a company goes down it goes *fast*, even for companies that were seen as the very image of solidity. Think of Lehman Brothers, the large financial company that went down in a few days at the time of the great financial crisis of 2008. That was when we discovered that there is no such a thing as a company that's "too big to fail".

While companies come and go, whole economies can experience disastrous crashes and, in that case, recovery may take a long time and sometimes never happens. Over history, economic collapses often accompany the decline and disappearance of empires and entire civilizations. Humankind has also seen abrupt population collapses caused by famine and pestilence and the same is true for the production of mineral resources that has seen entire regions experience production collapses, one of the most recent cases being that of the oil production from the North Sea. Today, we are facing the dire possibility of the ruin of our civilization and, perhaps, of the whole Earth's ecosystem. Climate change and resource depletion are the twin aspects of the troubles ahead.

Collapses are bad enough in themselves but they have a further quirk: they tend to arrive unexpected. Unless you are a firefighter, a physician, maybe you manage a large-scale electrical grid, or are engaged in some similar job, collapses are not part of your everyday planning. There is no "science of collapse" taught in universities or in business schools, and most of what we do is based on the idea that things will keep going on more or less as they have been doing in the past. The economy is supposed to be growing forever simply because it has been growing up to now. The same is true for the human population, the production of crude oil, or life expectancy at birth: they have been growing in the past and they are expected to keep growing in the future. The agencies and institutions that prepare forecasts in these fields work mainly on the basis of extrapolations of the historical data of the past few decades and tend to present a rosy picture of the future. It is a general problem we have with managing the future: nobody wants prophecies of doom! Yet, as we all know, growth cannot continue forever in a finite world (as we should know unless we are madmen or economists, a quote attributed to Kenneth Boulding). So, we should be prepared for the other side of growth well before collapse.

But what causes collapses? In ancient times, it seems that people tended to fault supernatural entities, Gods or evil magic, for the disasters befalling them. The first to note that collapses are a natural phenomenon, a fact of life, was perhaps the Roman Philosopher Lucius Annaeus Seneca in a note in on one of his letters to his friend Lucilius, written during the first century CE. Much later, during the seventeenth century, Galileo Galilei was the first scientist who tried to provide a mathematical explanation of collapses in the study of the fracture of solid objects.

Seneca's observation remained qualitative, while Galileo lacked the mathematical tools that he would have needed to build a complete theory of fracture. So, a true understanding of the physics of collapse came only in recent times with the development of the science of complex systems. The results of decades of work tell us that rapid changes are part of the way the universe works, a manifestation of the principle that rules everything, from living cells to galaxies: entropy, the basis of the second principle of thermo-dynamics. The science of complexity is possibly the most fascinating field of modern science and surely one that has significant consequences for our everyday life.

Out of this rapidly evolving field of science, there came the concept of the "Seneca Effect". It saw the light for the first time in 2011 in the form of a post in my "Cassandra's Legacy" blog [3]. Later on, I published a more detailed mathematical model in "Sustainability", a scientific journal [4]. Then, in 2017, I published a book that I titled "*The Seneca Effect*" [1]. I don't think that it is a difficult book to read, but it is also true that it was conceived as an academic book, with all the appropriate formulas and mathematical models. But the science of collapse is not just for academics: it is a science that everybody should know and use at least in its main features. That is the origin of this book, not a simplified version of the first Seneca book but a completely new one, with new examples, new discussions, new fields of application—also largely based on my personal experience.

So, this book is dedicated to how to confront collapses by being prepared before they arrive. That does not mean resisting collapse at all costs, desperately trying to maintain things as they are. Doing that means, typically, gaining some time in exchange for a much faster and abrupt collapse: by all means a bad deal. There are many examples of this concept and you can surely think of examples from your personal experience and perhaps the most evident one involves debating on social media. The more you try to contradict your opponent, the more you will find he or she will resist and respond to your arguments. That will often lead to the phenomenon called "flaming" that makes the discussion degenerate into an exchange of insults and personal attacks: a collapse of the debate!

Instead, the way to deal with collapses is to use what I call here the "Seneca Strategy". It is a view that derives from an interpretation of Seneca's work as a Stoic philosopher but that is also perfectly compatible with the modern field called "system dynamics" that Jay Forrester developed in the 1960s. The basic idea of the Seneca strategy is that the attempts to stave off collapse tend to worsen it [5]. It is also an idea with elements in common with some martial arts, such as Jiu-Jitsu or its modern incarnation, Judo, where practitioners aim at manipulating the opponent's force against themselves rather than confronting it with one's own force. So, the Seneca strategy consists in not opposing the tendency of the system to go in a certain direction but steering it in such a way that the collapse need not occur. The key of the strategy is to avoid that the system accumulates so much strain that then it is forced to vent it in an abrupt manner. Think of the story of the straw that broke the camel's back: collapse would not have occurred if the owner of the camel had avoided to overload the poor beast with heavy stuff.

But it is not always possible to avoid collapse, even though you may be able to detect it before it comes. Sometimes, it is just too late: the system has grown beyond its limits and it is now hovering somewhere in the unstable condition we call "overshoot". In this case, the system has to return to its acceptable limits, a condition sometimes called "carrying capacity". The best you can do is to soften the impact and prepare for landing. You will go through what I call the "Seneca bottleneck" with a view of restarting afterward and doing something better and wiser. That may be called the "Seneca Rebound". A good example is the fossil fuel industry: we can see its impending collapse and we *want* it to collapse in order to avoid a climate catastrophe, but not so fast that its fall will kill billions of people by depriving them of the energy they need to survive. The oil industry must keep extracting just the minimum that will be needed in order to create the renewable energy infrastructure that will replace the fossil one after the unavoidable collapse. This is what I called the "Sower's Way" [6], and it is a variant of the Seneca strategy.

Another useful skill derived from the Seneca strategy is how collapse can be exploited to get rid of old and useless structures, and organizations. I am sure that you know plenty of examples of irredeemably twisted and corrupt organizations that you have been thinking should be erased and rebuilt from scratch. You probably have in mind your government, but it is also possible to think of much smaller systems: plenty of people try to keep their marriage together beyond what's reasonable to do and in many cases divorce, the collapse of a marriage, is the best option. But a company may also become unfit to survive in the market, burdened by obsolete products, outdated strategy, an unmanageable organization. Bankruptcy is the way we call collapse in this case and, again, it is a way to start again from scratch. There are many other cases of collapses that result in something new and better emerging from the ashes of the old.

Finally, there is a further application of the science of collapse, one that Seneca would surely have disapproved of but that I cannot avoid mentioning: destroying one's enemy or competitor. It can be a military strategy: normally, a conflict ends when one of the two sides collapses and is not able to keep fighting any longer. It may happen because its military apparatus has been damaged beyond its resistance limits during the conflict but also as the result of the dark and dire things that, today, go under the name of "psyops" (psychological operations). Then, of course, nothing prevents people from using similar methods in business to cause the collapse of a competitor: think of "dumping", also defined as "predatory pricing". And even in love, perhaps the most competitive human enterprise, there exist objectionable but effective ways to get rid of competitors. Do you remember Hamlet saying, "Be thou as chaste as ice, as pure as snow, thou shalt not escape calumny"?

To summarize, the basis of the Seneca strategy can be described in four main points,

1. *Attention.* Remember that collapses occur and they do not just strike other people: they may strike you. Prepare in advance for a possible collapse!
2. *Avoidance.* You can avoid collapse if you start early enough by acting on the elements that put the system under stress. Detect collapses before they come!
3. *Mitigation.* If it is too late to avoid collapse, you can still reduce its damaging effects if you take appropriate precautions. Don't try to avoid collapse at all costs, but you can always soften it!

4. *Exploitation*. In some cases, you can use collapse to get rid of obsolete structures or to damage your competitors. And, therefore, welcome collapse!

I hope you will find this book useful for your life and your career but note that it is more than a manual for managing collapses. Since it starts from a sentence of a Stoic philosopher, it has a certain approach based on Stoic philosophy. The Stoics had understood a lot of things already two thousand years ago, the main one being, perhaps, that you cannot predict the future, but you can be prepared for it.

Firenze, Italy   Ugo Bardi

References

1. Bardi, U.: The Seneca Effect. Why Growth Is Slow but Collapse Is Rapid. Springer (2017)
2. Griffith, E.: A unicorn lost in the Valley, evernote blows up the 'Fail Fast' gospel. The New York Times (2019). https://www.nytimes.com/2019/06/28/business/evernote-what-happened.html. Accessed 28 Mar 2019
3. Bardi, U.: The Seneca effect: why decline is faster than growth. Cassandra's Legacy (2011). https://cassandralegacy.blogspot.com/2011/08/seneca-effect-origins-of-collapse.html. Accessed 7 Feb 2019
4. Bardi, U.: Mind sized world models. Sustain. **5**, 896–911 (2013)
5. Meadows, D.H.: Leverage Points: Places to Intervene in a System. donellameadows.org (1999). http://leadership-for-change.southernafricatrust.org/downloads/session_2_module_2/Leverage-Points-Places-to-Intervene-in-a-System.pdf
6. Sgouridis, S., Bardi, U., Csala, D.: The sower's way. Quantifying the narrowing net-energy pathways to a global energy transition (2016). arXiv:1602.01203

Acknowledgements

The first person to be thanked for this book is Charles Hall, pioneer of biophysical economics, who set so many things in motion in a field that would not have been the same if he had not been around. Dennis Meadows and Jorgen Randers are among those who set me on the path of studying world dynamics, just as Colin Campbell taught me about oil depletion. Then, Dave Packer who made this book possible, Dmitry Orlov for having pushed me to develop a model for fast crashing systems, Luca Mercalli for reminding me of how Seneca had described ruin in his letters. Thomas Gaudaire-Thor for telling me about the fascinating concept of "egregore". My coworkers Francesca di Patti, Sara Falsini, Gianluca Martelloni, and Ilaria Perissi have been working with me in the difficult field of modeling the world's economy. My Russian coworkers and friends, Konstantin Eltsov, Andrey Klimov, Irina Kurzina, and Tatiana Yugay, among many others, showed me how to survive a bad societal collapse. My students Beatrice Barletti, Ilaria Garbari, Federico Licciardi, and Koi Leo Ian Ioshi Merc withstood my rambling lectures and kindly accepted to pose for one of the illustrations of the present book. I would also like to thank Susan Kucera for her interest in these matters that she transmits in her movies. Alessia Roberta Scopece, classicist and medievalist, has been a remarkable source of inspiration, while Valeria Fenudi provided some of the figures of the book, Miguel Martinez edited and reviewed the text. Then, my wife Grazia was kind enough to bear with me during my full-immersion days in writing this book. And all the readers of the blog "Cassandra's Legacy" who kindly suggested a list of titles that were useful to select what I think is the right one. Finally, a note of praise for my fictional single-celled assistant, Amelia the Amoeba, who turned out to be quite popular with my students.

Contents

1

The Science of Doom: Modeling the Future

Forecasts are not always wrong; more often than not, they can be reasonably accurate. And that is what makes them so dangerous. They are usually constructed on the assumption that tomorrow's world will be much like today's. They often work because the world does not always change. But sooner or later forecasts will fail when they are needed most: in anticipating major shifts in the business environment that make whole strategies obsolete.
—Pierre Wack [1]

I will not die one minute before God has decided.
—Mike Ruppert, *Crossing the Rubicon* [2]

Predicting the Future: The Russian Roulette

Fig. 1.1 The Author giving a talk in Florence, in 2018. Note the gun in his hand: it is a harmless toy used to focus the attention of the public on the fact that knowledge, or lack thereof, may be dangerous. This may happen with guns, but also with much larger entities such as climate change (*photo courtesy* Ilaria Perissi)

© Springer Nature Switzerland AG 2020
U. Bardi, *Before the Collapse*,
https://doi.org/10.1007/978-3-030-29038-2_1

When I give public talks, sometimes I take a toy gun with me and I show it to the audience. I ask them this question: imagine you had never seen a gun, how would you know what it is and what it is for? Usually, the people in the audience immediately understand the message: the gun is a metaphor for climate change. How do we know how the Earth's climate works? And how can we know the kind of damage it can do to us? It is all about the field we call "epistemology," how do we know the things we know? Whether we deal with firearms or with climate change, ignorance can kill and epistemology can be a tool for survival (Fig. 1.1).

The idea of an unknown artifact that turns out to be a weapon is a typical trope of science fiction. When the hero of the story happens to find a ray gun or a phaser left around by aliens, he usually manages to understand immediately what it is for and to use it against his extraterrestrial enemies, it is a theme seen recently in the movie "Cowboys and Aliens" (2011). Rarer is the case of aliens stumbling onto a human-made weapon, but the theme was explored by Gilda Musa [3] in a delicate and intelligent story written in the 1960s where human explorers introduce a handgun to a civilization of peaceful aliens. Tragedy ensues, as you may imagine.

So, let us follow this idea. Suppose you are an alien and that, somehow, you find this strange object. You have never seen anything like it before and you only know that it was left by those weird Earthlings. They are a tricky race, so you may suspect that it is a dangerous object—maybe a weapon. But how to tell? Framed in these terms, we have a very pragmatic question that does not lead us to ethereal philosophical reasoning. What we need to do is to build a *model* of the unknown entity that can tell us how to deal with it and—in particular—if it is dangerous or not to deal with it.

Some people tend to belittle models as something purely theoretical, as opposed to the real world. But that's a completely wrong view: models are necessary and we build them all the time in our everyday life. On this point, it is worth citing Jay Forrester, one of the greatest model builders of the 20th century, the person who developed the method of calculation used for "*The Limits to Growth*" study [4].

Each of us uses models constantly. Every person in private life and in business instinctively uses models for decision making. The mental images in one's head about one's surroundings are models. One's head does not contain real families, businesses, cities, governments, or countries. One uses selected concepts and relationships to represent real systems. A mental image is a model. All decisions are taken on the basis of models. All laws are passed on the basis of models. All executive actions are taken on the basis of models. The question is not to use or ignore models. The question is only a choice among alternative models.

Models can be complicated or simple, they may be based on equations, analogies, or just intuition. But they are always the same thing: entities existing in our minds that help us plan ahead and avoid the many disasters that could await us. Models are often useful, especially if they are tested by experience, but may also be disastrously wrong. Returning to the example of the gun as an unknown object, there are various ways you can make bad (actually, deadly) models about it. For instance, you know of the "Russian Roulette" game. It involves loading the cylinder of a revolver with a single live round, spinning it at random, and then pulling the trigger while the barrel is pointing at one's head. The origins of this game (if we want to define it in this way) are fictional—its first mention goes back to a novel by the Russian writer Lermontov *Hero of Our Time* (1840). But some people do play the game for real. We don't have good statistical data but a 2008 paper [5] reports 24 cases of Russian Roulette deaths in Kentucky from 1993 to 2002. Extrapolating these data to the whole US, we could roughly estimate that every year around 10–20 people die of the Russian Roulette and, possibly, a hundred or so play it and survive.

For most of us, it is evident that the only way to win at the Russian Roulette is by not playing it but, evidently, some people have a wrong understanding of statistics and use it to make very bad models. That must be not so uncommon, otherwise nobody would ever play any roulette game, not just the Russian version with a gun. But people do engage in gambling, sometimes using dangerous strategies such as the "Martingale" that nearly guarantees disastrous losses [6]. Compulsive gamblers face sometimes the same kind of Seneca ruin that the Russian roulette can generate but, in their case, the cliff may start from one of the windows of an upper floor of the casino building [7].

Some people, apparently, tend to see the world as dominated by forces that cannot be quantified in statistical terms. They seem to believe that, if your destiny is decided by God's plan, there follows that the Russian Roulette cannot kill you: you will die only if He decides that you have to, otherwise you will live. Of course, few people trust God to the point that they risk shooting themselves: after all, God is supposed to be benevolent and merciful but His patience is also known not to be infinite. Nevertheless, it is not rare to encounter a similar attitude in discussions on climate change. Some people are so convinced that the Earth's climate is in the hands of God that it is evident for them that nothing mere humans do can alter it, surely not by increasing the concentration of a greenhouse gas of a few hundred parts per million. And, for this reason, humankind seems to be engaged in playing a deadly game of Russian Roulette with the Earth's climate as a loaded gun.

Can we build better models than these examples that put our life at risk? Of course, we can. Generally speaking, there are two ways to build models: the "top-down" approach and the "bottom-up" one. The top-down method is sometimes based on statistical data and consists in treating the system as a "black box." You look at what the system does and you build up a model on the basis of what you see, without worrying too much about the inner mechanisms of what you are examining. A modern version of this heuristic approach is called the "Bayesian Inference Method." The idea is that you first assign a certain probability to the hypothesis (this is called the "prior",) then you update it to a new value (called the "posterior") in the light of new data or evidence. Then you iterate, until a certain stable value is obtained or, in any case, adapt your estimates to a changing system. This is a variant of the general "heuristic" model of using statistical data to predict the future.

The other method, the bottom-up one, is sometimes called the "reductionist" approach and is the basis of the scientific method. It consists in separating the system into subsystems and examining each of them separately, then building a model of how the whole system works. As you know, this method is relatively new in human history. It was formalized in the way we know it only a few centuries ago and is still being tested and refined.

Both methods have limits. In particular, they require specialists and appropriate tools for a thorough examination that is expected to provide a complete understanding of the system you are studying. And that also requires time while, in the real world, often you have neither the resources nor the time needed to apply these methods in full. Especially when dealing with things that could be dangerous, you cannot wait to have scientific certainty, assuming you can ever have it.

In particular, the statistical inference method, also in its Bayesian version, can lead you to dangerously wrong models. A classic mistake here is "the law of small numbers," identified for the first time by Twersky and Kahneman in

1971 [8]. The law says most people tend to build models on the basis of too few data. In particular they may engage in (1) gambling on the basis of small samples without realizing the odds, (2) having undue confidence in early trends and in the stability of observed patterns, (3) having unreasonably high expectations about the replicability of significant results, and (4) always finding a causal "explanation" for any discrepancy.

Let us apply the law of small numbers to the example of the gun. Assume you are one of the aliens of Gilda Musa's story and that you are tinkering with the strange object left by Earthlings, trying to understand how it works. You note the presence of a metal thing that looks like a lever. You test it by pulling it with your finger and, yes, it is a lever: it acts on the cylinder, making it spin. It seems to be a trigger that acts on another small object on the opposite side of the cylinder: it goes up and down, making a clicking noise. You pull the trigger a few times and the result is always the same: nothing more than that clicking sound: maybe it is a musical instrument? The Bayesian inference method tells you that the probability of the object being a weapon goes down every time that you pull the trigger and nothing happens. At the same time, the hypothesis that the object is a musical instrument becomes more and more probable. Then, to hear the clicking sound better, you place the object close to your head—the barrel-like protrusion on one side directly touching your ear. You pull the little lever once more and…

We can clearly see the problem of small numbers at work, here. Testing a revolver just a few times cannot tell you if there is a live round in one of the chambers of the cylinder. And a devilish result of the Bayesian inference process is that the more times you try and nothing happens, the more likely it seems to you that the object is harmless. The problem is there also with such things as climate change, oil depletion, resource depletion, poisoning of the biosphere, and more. We do have data for these systems, but often not for sufficiently long time spans: for instance, climate change is a very slow process that may turn out to be disastrous, but only in a relatively remote future. So, there arises the idea that since nothing horrible has happened to us so far, it never will—it is a wrong application of the Bayesian method. One of its forms is, "people have been saying that crude oil would run out on some already past date. That didn't happen and there follows that oil is not going to run out in the future." And, as you know, the words "so far, so good" were the last ones pronounced by the guy who was falling from the 20th floor of a building.

A good example of the limitations of heuristic methods when used alone can be found in the debate about "The Limits to Growth", (1972) [9], a study that attempted to describe the evolution of the world's economy. It was not a heuristic model: it did not treat the world's economy as a black box. The authors disassembled the economic machine taking into account the available natural resources, the effect of pollution, the growth of the human population, and more. This approach turned out to be incomprehensible to many economists trained in the statistical approach called "econometrics," a set of techniques used to derive a model directly from the historical data. In the well-known textbook by Samuelson and Nordhaus, *Economics*, (published for the first time in 1948 by Samuelson alone) econometrics is described as a tool "to sift through mountains of data to extract simple relationships."

On the basis of this approach, in 1972, William Nordhaus, who would later obtain the Nobel prize in economics, published a paper titled "Measurements without data" [10] where he harshly criticized the approach of "The Limits to Growth" study (even though he actually targeted an earlier, similar study by Forrester [11]). Nordhaus stated that the model:

> …..contains 43 variables connected to 22 non-linear (and several linear) relationships. *Not a single relationship or variable is drawn from actual data or empirical studies.* (emphasis in the original)

Note how Nordhaus is thinking in terms of econometrics, that is, one should extract relationships from the data rather than use physical considerations. It was the start of a degeneration of the debate that veered into a clash of absolutes and eventually consigned the "Limits to Growth" report to the dustbin of the wrong scientific theories from which it is only now slowly re-emerging. It is a story that I told in detail in my 2014 book "*The Limits to Growth Revisited*" [12].

The clash was created by a deep epistemological divide between two different approaches. In his papers, Nordhaus contrasted the "Limits to Growth" model with a model of his own [13] that he had developed on the basis of an earlier model by Solow [14], based on the fitting of the previous trends of the economy. It was a nearly completely heuristic model: it was based mainly on past data, and, since no collapse had taken place during the period considered, the model could not and did not foresee a collapse. Nearly 50 years after the debate, we can say that both Nordhaus' model and the "base case" scenario of

The Limits To Growth were able to describe the trajectory of the world's economy with reasonable approximation [15]. The two models diverge with the third decade of the 20th century and the optimism of Nordhaus and other economists could turn out to have been another case of the mistake that comes from the law of small numbers described by Twersky and Kahneman.

In general, the emphasis on only looking at data without even trying to build physical models can be seen as related to the approach called "Zetetics" [16] from a Greek word meaning "I search." Zetetics is an extreme form of the experimental method: zeteticists assume that data are all the need to understand the world. The term "zetetic" is often applied to the modern "flat-earth" movement whose adherents seem to think that since the Earth looks flat, then it must be flat. They refuse to see the Earth as a sphere because the evidence for a spherical shape is a theory, not a direct experimental observation. As a method of inquiry, zetetics may have some good points but, if it is applied in a literal manner, it can be suicidal. In the example of the gun, zeteticists would refuse to believe that a gun can kill anyone until they saw it actually killing someone and, possibly, they would maintain that this proves only that the specific gun having been tested is dangerous. On a much larger scale, the zeteticist's position "bring me experimental proof" could lead the whole humankind to an apocalyptic disaster caused by the consequences of climate change (but it must be said that Flat-Earthers, to their honor, do think that human-caused climate change is real [17]).

So, just looking at statistical data can easily lead us astray with models of complex and potentially dangerous systems such as the Earth's climate. How about the other possible method, the "bottom-up," reductionist approach? Is it better at making good models than the statistical approach? In some cases, yes, and, indeed, it is the basic tool of the "hard" scientific method. In fields such as physics and chemistry, scientists are used to performing carefully contrived laboratory experiments where they separate and quantify the various elements of systems that may be very complex. In engineering, for instance, the capability of a certain element of a structure, say, a plane or a bridge, is studied by performing separate tests on the materials that compose it. It is assumed that the behavior of a metallic alloy in the form of an hourglass specimen in a testing machine will be the same as in a real structure. Normally, that turns out to be correct, even though it is a conclusion that has to be taken with plenty of caution.

Applying the reductionist model to the example of the gun as an unknown object implies dismantling it. The experimenter should be able to determine that the object that goes up and down, pushed by the lever at the bottom, can hit and ignite the chemicals contained inside a small brass cylinder which, in turn, would propel out of the object a chunk of a few grams of lead at a speed of a few hundred meters per second. By all means, the reductionist method can tell us that this thing is *very* dangerous.

Within some limits, the reductionist approach is possible also for more complex systems, for instance the Earth's climate. We can identify several subsystems of the Earth's atmosphere, then study each one separately. The fact that carbon dioxide (CO_2) absorbs infrared radiation has been known since the early experiments by John Tyndall in 1859. Then, in 1896, Svante Arrhenius was the first to propose that CO_2 had a warming effect on the Earth's atmosphere and that the burning of fossil fuels would cause an increase of the atmospheric temperatures [18]. It was the origin of the idea of global warming caused by the effect of "greenhouse gases" and the "greenhouse effect," even though Arrhenius did not use these terms. Over the years, more and more sophisticated models were developed to tell us what kind of temperature increase we can expect if we continue to dump greenhouse gases into the atmosphere.

But, of course, neither Arrhenius nor anyone else could make a laboratory experiment proving the concept of greenhouse warming of the Earth's atmosphere. Some enthusiastic amateurs try to do just that at home using glass jars or Coca Cola bottles. Most of these experiments turn out to be poorly made or simply wrong [19]. Even when they are done correctly, all they can show is that an irradiated glass vessel gets a little warmer when it contains more CO_2 inside. But that proves nothing more than what Tyndall had already demonstrated one and a half centuries ago. The problem is that the properties of the atmosphere cannot be exactly reproduced in a laboratory: just think of the variable density of the atmosphere as a function of height, you cannot reproduce that in a Coca Cola bottle!

It is a problem that's especially acute with some models of the atmosphere. You may have heard of the "biotic pump" theory developed by two Russian researchers, Victor Gorshkov and Anastassia Makarieva [20]. The theory aims to explain the fact that rainforests manage to attract a high amount of rainfall and is based on a physical phenomenon, that when water vapor condenses it creates a negative pressure. The idea is that the biotic pump keeps the forest wet by continuously pumping moisture from the oceans. It is a fascinating theory but how can we prove it is correct? You can't create a rainforest in a lab and the only way to test the theory is by means of model-building and comparison with real-world data. It will take time before an agreement on the validity of this theory will be reached by the scientific community.

Does that mean that the idea of human-caused global warming is not supported by experimental data? Not at all, but you must understand how the scientific method deals with this kind of systems. The basic physics is known, the parameters of the system can be measured, the interaction among parameters can be simulated in computer models and that is enough to arrive at a number of well-known conclusions, such as that, at present, CO_2 is the main driver of the observed warming of the Earth's atmosphere.

As we all know, not everybody accepts this conclusion. In most cases, the denial of the basic features of the global warming phenomenon is based on purely political considerations. Some people state that the whole story is a hoax created by a cabal of evil scientists who wanted more money for themselves in the form of research grants. Of course, it is not possible to rigorously prove that this is not the case, even though it may be reasonably argued that the existence of such a cabal is, at best, a highly unlikely assumption. But, sometimes, denial is based on a zetetic approach: it is often claimed, for instance, that there is "no proof" that CO_2 warms the Earth. In this kind of epistemological approach, in order to "prove" that CO_2 warms the Earth, you would need a controlled series of experiments where you control the concentration of CO_2 in the atmosphere while measuring the effects on temperatures and also where you check the effects on the planetary ecosystem. An experiment to be done at a planetary scale and, obviously, a little difficult to do, especially for the part that involves the collapse of the ecosystem.

Overall, we can say that there are many ways to see the world but that none gives us absolute certainty of what the future could be. We always try to do our best, but we are not always successful. Sometimes we err because of an excess of caution, in others because we are careless or overoptimistic. Nevertheless, it is a good idea to use models to understand the world around us and build models for what we expect from it. The scientific method, while not a panacea, can help us a lot in the task. Trusting God may also help but, as the old saying goes, try to keep your powder dry.

How Good Can a Model Be? Nightfall on Lagash

Fig. 1.2 A mechanical planetarium ("Orrery") made by Benjamin Martin in London in 1766, presently at the Putnam Gallery in the Harvard Science Center. This mechanical model is possible because the solar system is not a complex system and the planetary orbits are stable and exactly predictable (Figure courtesy of Sage Ross. https://en.wikipedia.org/wiki/Orrery#/media/File:Planetarium_in_Putnam_Gallery_2,_2009-11-24.jpg)

In 1941, Isaac Asimov published one of the best-known science fiction stories of all time, "*Nightfall.*" It told of a remote planet called "Lagash," inhabited by a species of intelligent aliens. In the story, Lagash is constantly illuminated by at least one of the six suns of its multiple star system but, every some thousand years, an eclipse of the main sun causes the side of the planet where the Lagashians live to fall into complete darkness. They are completely unprepared for sudden darkness, the shock causes them to go mad and they start burning everything at hand, just to have some light. That is the cause of the cyclical collapses of their civilization that Lagashian archaeologists had noted but had been unable to explain.

The drama in Asimov's story is related to how a group of Lagashian scientists has been able to predict the coming nightfall by studying the motions of the suns of the system and then extrapolating their trajectories. Here is how the prediction is told by one of the scientists in the novel,

The complex motions of the six suns were recorded and analyzed and unwoven. Theory after theory was advanced and checked and counterchecked and modified and abandoned and revived and converted to something else. It was a devil of a job. <..> It was twenty years ago that it was finally demonstrated that the Law of Universal Gravitation accounted exactly for the orbital motions of the six suns. It was a great triumph.

Here, Asimov tells us how the so-called "hard" sciences, physics in particular, can provide models whose predictions are *exact*. The Lagashians had a hard time in finding the law of universal gravitation because their star system was much more complex than the Solar System, where planets describe nearly circular orbits around a single sun. But eventually they arrived at the same result reached by their terrestrial colleagues and they were able to use the law to make predictions. Asimov was a scientist himself and his stories were based on solid physics. In 2014, Deshmuk and Murty carried out calculations to show that a star system similar to the one described by Asimov could actually exist [21].

Leaving aside the complicated star system of Lagash, here on Earth we know very well that Newton's gravitation law is one of the strong points of classical physics, to the point that the prediction of eclipses is one of the most impressive successes of astronomy. So much that a trope of novels and movies is how a stranded explorer impresses the ignorant people of some remote tribe by predicting a solar eclipse that, later on, punctually takes place. Then, the tribesmen make him their Godking or something like that. The Solar System is truly a clockwork and the movements of the major bodies that are part of it are regular and predictable. Indeed, during the 18th century, mechanical models of the Solar System based on the technology of clocks became fashionable. These models were called "orreries" from the name of Charles Boyle, 4th Earl of Orrery (Fig. 1.2).

But how precise can a model be? In some cases, it can be very precise. You can use Newton's law to calculate the motion of a space probe and direct it towards a destination hundreds of millions of kilometers away from the Earth. In principle, you can use the law to calculate the trajectory of any chunk of mass in motion in a gravitational field. Maybe you could do that for every atom moving in the universe: it would be just a question of knowing what forces act on it and what is the current speed and position of each particle. Then you would apply Newton's gravity equation, also taking into account electric and magnetic fields, and in this way you could predict exactly the trajectory of all the particles in the universe. You would have an all-powerful model telling you exactly what the future will be.

This view is called "scientific determinism" and is normally attributed to Pierre-Simon de Laplace (1749–1827). In his *A Philosophical Essay on Probabilities* (1814), he spoke of an "intelligence" able to have this kind of knowledge that would make her/him/it all-powerful. Later on, the term "Laplace's demon" was coined for this hypothetical creature (but why not a demoness? Gender correctness should impose that). Clearly, if you had the computing power to simulate the demoness, you could predict the future with great precision: no collapse would escape advance detection. Just like the fictional astronomers of Lagash were able to predict the solar eclipse that would throw their world into chaos, we would be able to predict such things as earthquakes and hurricanes. Even financial crises would be detected well in advance: economic agents are made of atoms, too!

I don't have to tell you that this is not possible. There exist good scientific reasons, quantum mechanics, thermodynamics, chaos theory, and more, telling us that Laplace's demoness would rapidly get confused and would lose her way through the galaxies. But, without going into these matters, there are simple practical problems that make exact long-term predictions impossible. Richard Feynman discusses this point his book "*Lectures in Physics*" (1964) (pp. 2–9 of the third volume):

> It is true, classically, that if we knew the position and the velocity of every particle in the world, or in a box of gas, we could predict exactly what would happen. And therefore the classical world is deterministic. Suppose, however, that we have a finite accuracy and do not know *exactly* where just one atom is, say to one part in a billion. Then, as it goes along it hits another atom, and because we didn't know the position better than one part in a billion, we find an even larger error in the position after the collision. And that is amplified, of course, in the next collision, so that if we start with only a tiny error it rapidly magnifies to a very great uncertainty. … given an arbitrary accuracy, no matter how precise, one can find a time long enough that we cannot make predictions valid for that long a time.

So, all measurements suffer from uncertainties and these uncertainties tend to accumulate, becoming larger as time goes by. Eventually, the uncertainty becomes too large to make any prediction possible. For instance, a good mechanical chronometer may have an accuracy of the order of 5 seconds per day, so it can keep telling an approximately correct time for several days—even several weeks if you are not too fussy. But not for several months, unless you synchronize it periodically with some other, more accurate, timekeeping device. If you can't do that, it is like in the old joke that says that the best watch is the watch that has stopped because, at least, it can tell the exact time twice a day!

Even for the Solar System, we cannot exclude that, hundreds of millions of years in the future, the interactions among the several bodies of the system could lead to destabilizing the orbits of some of the planets, maybe having Mercury, Mars, or Venus colliding with each other or with the Earth. The possibility that the system is chaotic over a long time scale, that is it follows an unpredictable and always different trajectory, has been discussed several times and some scientists claim that it is the case [22]. So, it is not impossible that in the future the solar system might go through some kind of a Seneca Cliff with the planetary orbits being destabilized by internal oscillations. Or, alternatively, another star could glide nearby, as happens in H. G. Wells' short story "*The Star*" (1897), one of the first pieces of fiction to describe a cosmic catastrophe. In the story, the star moves away after having caused all sorts of disasters for humankind but leaving no permanent damage to anything. However, were this to happen for real, the gravity of the star might affect the system strongly enough that the planets would whirl away from their orbits. Fortunately, the density of stars in our region of the galaxy is low enough that the probability of this kind of collision is truly infinitesimal.

Leaving aside these dramatic scenarios, the limit of all models of the future is the gradual loss of information as we move forward in time. It is part of the general laws of the universe, it is entropy doing its work. A good example is the model called "random walk." Imagine that you see a drunken man standing somewhere on the sidewalk. You see him, so you know exactly where he is. Now, suppose you walk away, planning to come back after a while. In the meantime, the drunken man—being drunk—will be walking at random in one or the other direction along the sidewalk. The question is, can you predict where he will be when you come back?

In this simple case, you do not need equations to understand that if a step in one direction is as probable is as a step in the opposite direction, then the most probable point where to find the drunken man on the sidewalk will be the starting point. But it takes a mathematical treatment developed for the first time by Gauss to determine the probability of finding the drunkard at a certain distance from the starting point as a function of the number of steps. This distance increases gradually: the Gaussian model tells you that it goes with the square root of the number of steps.

These are not just considerations designed to make statisticians happy. They have applications in the real world: for instance, imagine you are commanding an anti-submarine ship and are searching for an enemy submarine, you know where it was at a certain moment but then you lost track of it. Now, where could it have gone? You could just search at random, as players do in the "Naval Battle" boardgame, but there are better ways. Statistical models can tell you something about where the submarine could be, assuming that the captain has been moving at random to escape detection. There are more sophisticated models, such as the "Lévy flight" (LF) one, where the searching agent moves randomly but with a "power-law" distribution of jumps. It means that the probability of a long-range jump is larger than it would be in a purely random search. This algorithm makes the searching force act like a wild predator that jumps around in search of the prey, as real predators seem to do when they have no target within their sensory range [23].

These considerations tell you something about the limitations of models: none can exactly predict the future. But that doesn't mean that models are useless. It is just that we have to use models knowing what they can do and what they cannot do. The future cannot be predicted, true, but that does not mean you cannot be *prepared* for the future, and that is what you really need. Not perfect and all-encompassing models, but good enough models. And most models can be good enough if you are careful to avoid asking them to do things they cannot do.

So, we need models and models can be simple or complicated, may be based on mathematical equations, physical laws, statistical inference, or simply human intuition. It is always the same concept: a model is a virtual representation of the real thing. We can "run" it in virtual space in our mind or using a computer and hope that the model will describe well enough the real system it is supposed to describe and tell us what we can expect from it.

Some models are plainly wrong, such as assuming that all blondes are dumb. Jay L. Zagorsky of the Ohio State University used a statistical analysis of actual data to check the idea [24]. The result was that, as you may have expected, blondes are no less intelligent than women with different hair color. And not only that, they might be slightly smarter, even though this result is reported to be statistically not significant. So, the study confirmed a statement by Dolly Parton (1946–), blonde American singer and actress, "I'm not dumb… and I'm not blonde, either." On the contrary, some models are so good that they can be deadly. If you are at war and a rain of artillery shells lands into your trench you should conclude that the enemy is using good models to guide their fire or, equivalently, that your side is using bad models. As every soldier knows, friendly fire never is such.

The basic rule in choosing models is that they have to consider all the relevant parameters. On this point, it may be worthwhile to cite the old joke of the girl who wanted a perfect marriage. She carefully organized every detail of the ceremony: the dress, the food, the cake, the flowers, the bridesmaids, the groomsmen, and everything else. She made only one mistake: the husband. But there is also the possibility of the opposite mistake: that the model includes too many parameters. Should the model try to follow reality in all its minute details, or can it neglect some? There are no firm rules on this point, the temptation for many modelers is to include as many parameters as possible—it makes the model look "better" and, for the layman, a complicated model often has a ring of truth that comes from the very mysterious way it works.

Sometimes, people approach these hugely complicated models with the same awe that in ancient times people must have felt when approaching the Pythoness of the Oracle of Delphi. But many parameters increase the uncertainty and, with a large number of degrees of freedom in the equations of the model, there is always the risk of being able to obtain an excellent fit of the data with the wrong model. In some cases, a good fit can be obtained for some sets of data, but the models have no physical basis and poor predictive capabilities. In the end, the problem was well described by John von Neumann when he said

> With four parameters I can fit an elephant, and with five I can make him wiggle his trunk.

The problem of too many parameters plagues many models and it can be mitigated by the procedure called *sensitivity analysis* [25], the study of how the output of the model is related to changes in the input parameters. It is an important concept that can test such things as the robustness of the model, that is whether small variations in the input lead to large changes in the output. It can also determine if some parameters have a negligible effect and can be neglected without losing the predictive ability of the model.

The problem of too many parameters can also be approached following the concepts developed by Seymour Papert in the 1960s, when he developed the "Logo" programming language [26]. The idea, in this case, was to use "mind sized" models. Papert was interested in promoting learning and his idea was to make models simple enough that they could be grasped and understood by an average human mind. It is a good approach, the problem is that for most people "simple" means "inaccurate" and it is difficult to convince politicians and decision makers that you can say something about where the world is going by means of just a few parameters.

Supposing that you have a model, then you must know how to use it. Having a hammer in hand notoriously leads people to believe that everything is a nail and you should avoid this mistake with models. They are tools and you must know how to use them and for what purpose. Do you really want to predict the future? And, if so, with what kind of accuracy? And, more importantly, how will you react to the predictions that the model makes? The last question is probably the most important one about models. Models are often used for forecasting, but forecasting is often wrong. So, how should people deal with the uncertainties inherent with models?

This issue was clearly identified by Pierre Wack—strategic analyst at shell oil—when he separated two fundamentally different categories: forecasts and scenarios. In 1985 he wrote [1]

> It is fashionable to downplay and even denigrate the usefulness of economic forecasting. The reason is obvious: forecasters seem to be more often wrong than right. Yet most US. companies continue to use a variety of forecasting techniques because no one has apparently developed a better way to deal with the future's economic uncertainty.
>
> …
>
> Few companies today would say they are happy with the way they plan for an increasingly fluid and turbulent business environment. Traditional planning was based on forecasts, which worked reasonably well in the relatively stable 1950s and 1960s. Since the early 1970s, however, forecasting errors have become more frequent and occasionally of dramatic and unprecedented magnitude.
>
> …
>
> Most managers know from experience how inaccurate forecasts can be. On this point, there is probably a large consensus.
>
> My thesis—on which agreement may be less general—is this: the way to solve this problem is not to look for better forecasts by perfecting techniques or hiring more or better forecasters. Too many forces work against the possibility of getting *the* right forecast. The future is no longer stable; it has become a moving target. No single "right" projection can be deduced from past behavior.
>
> The better approach, I believe, is to accept uncertainty, try to understand it, and make it part of our reasoning. Uncertainty today is not just an occasional, temporary deviation from a reasonable predictability; it is a basic structural feature of the business environment. The method used to think about and plan for the future must be made appropriate to a changed business environment.

According to Wack, the best way to deal with the future is not by means of forecasts but by means of *scenarios*. Scenarios are what the military use to plan for war, sometimes called wargames. In war, uncertainty is a basic feature of the situation: if it were possible to forecast in advance the outcome of a war there would be no need to fight it. In business or in other sectors, scenarios are the same thing: a fan of possibilities that we should be prepared for. According to Wack [1],

> Even good scenarios are not enough. To be effective, they must involve top and middle managers in understanding the changing business environment more intimately than they would in the traditional planning process. Scenarios help managers structure uncertainty when (1) they are based on a sound analysis of reality, and (2) they change the decision makers' assumptions about how the world works and compel them to reorganize their mental model of reality. This process entails much more than simply designing good scenarios. A willingness to face uncertainty and to understand the forces driving it requires an almost revolutionary transformation in a large organization. This transformation process is as important as the development of the scenarios themselves.

And we see what is the problem: Wack was writing in 1985 about work he had been doing in the 1960s. Things have not changed much from them: companies still pay people to produce forecasts which regularly turn out to be wrong, an especially sad story is that of the attempts at predicting the prices of crude oil. You can amuse yourself looking for old forecasts and be bemused at discovering how bad they can be. For instance, nobody had ever predicted the great spike of oil prices of 2008 that brought the value of the oil barrel to the all-time high of about $150. And nobody had predicted the price collapse that would follow. But Wack and his team had correctly evidenced that "something" was going to happen around 1970, even though they could not exactly predict what.

Wack was operating accordingly to a concept that later on, in 2007 Nassim Taleb would define the "turkey fallacy" in his 2007 book "*The Black Swan,*" [27] where he says:

> Consider a turkey that is fed every day. Every single feeding will firm up the bird's belief that it is the general rule of life to be fed every day by friendly members of the human race "looking out for its best interests," as a politician would say. On the afternoon of the Wednesday before Thanksgiving, something *unexpected* will happen to the turkey.

Note how here we do not have a problem of a too small sample. The turkey of the story may have hundreds of data points, up to 364, and believe that they are sufficient to show that humans are indeed a benevolent race dedicated to the well-being of turkeys. Every day that the turkey is fed this belief is reinforced. The same problem exists with climate change where the fact that the change is very slow leads people to a false sense of safety. "Temperature has been increasing a little," some people say, "sure, and so what? Nothing bad has happened to us, so far, so why should that change?"

The problem, in these cases, has a name: "tipping point" [28]—it is something typical of those systems called "complex" which tend to switch rapidly from one condition to another. We will see more about these systems in the next section but, for the time being, it is enough to note that they are very common in the real world. Think of the turkey of the story, it can have two distinct and separate states: a live turkey and a dead turkey. The same two states may occur for the whole of humankind when facing climate change.

Nassim Taleb uses the term "gray swan" for extreme events or conditions that were never experienced before but that might have been statistically predicted. A good example of gray swan is the Tōhoku tsunami that hit Japan in 2011. The tsunami wave was so large that it overcame the coastal defenses built on the hypothesis that such an event was too unlikely to be considered, yet it was not impossible on the basis of the known data on the sizes of historical tsunamis. A different case is that of events totally outside the statistical distribution, true black swans or "dragon kings" as they have been termed by Sornette [29]. In these cases, the system behaves in ways just not predictable from previous historical data. No matter how many times you test a gun by having the hammer hit an empty chamber, you won't have data about what happens when the hammer hits a live round. The transition between one state and another fits well the concept of the "Seneca Cliff" we are discussing in this book and these considerations apply also to climate predictions. In the heat of the debate, there is a point that the critics of climate science nearly always miss and that sometimes is missed also by supporters. The problem is that the current models are limited in terms of their capability of predicting extreme events. That is, they are not made to foresee the possibility of rapid, unexpected, and catastrophic variations, the concept described here in terms of the "Seneca Effect."

We can think of plenty of ways that models cannot describe for global warming to become dangerous for humankind, actually deadly. Rising temperatures could lead to the collapse of large fractions of the Greenland and

Antarctica ice sheets and that would generate truly catastrophic rises of several meters, even tens of meters, in the average sea levels. Then, of course, there is the "big one" in climate change: the tipping point that could irreversibly propel the Earth's climate system to the condition known as "Hothouse Earth" (sometimes "Greenhouse Earth") with average temperatures of some 5–8 °C higher than the present ones, no icecaps, no continental glaciers, a sea level rise of tens of meters, partial deoxygenation of the atmosphere. In the past, these conditions led to huge mass extinctions and, in some cases, to the near death of the whole ecosphere. The climatologist James Hansen even hinted at the possibility of a runaway greenhouse effect [30] that could lead the Earth to become a planet similar to Venus: with no life, an atmosphere composed mainly of CO_2 and sulfuric acid, and temperatures in the hundreds of degrees °C (now *that* would be a true Seneca Cliff!). Fortunately, the current knowledge of the physics of the Earth's atmosphere indicates that the Venus scenario is unlikely, perhaps impossible [31]. But the mere fact that these possibilities exist shows that we could be playing Russian Roulette with the Earth's climate.

All that is well known in the scientific debate but catastrophic tipping points are conspicuously missing in the political debate. The risk is sometimes hinted at but never given full attention, probably because scientists have been afraid of being branded as catastrophists if they were to make their worries public. They seem to have been practicing a kind of self-censorship that makes them avoid stating their worst worries in public [32]. At least so far, the whole discussion seems to be governed by this kind of self-censorship and we do not know what could happen if the concept of climate catastrophe were to gain the center of the debate. James Schlesinger is reported to have said that "people have only two modes of operation: complacency and panic" and it is not obvious that moving the discussion to panic mode would lead to the best choices in order to avoid a climate disaster. How we will manage to deal with the problem, if ever we will, is all to be seen.

Overall, we can apply to models Shakespeare's quote from Hamlet, "Be thou as chaste as ice, as pure as snow, thou shalt not escape calumny." We only need to change a few terms and we obtain: "Be thou as precise as the 5th decimal place and as accurate as better than 1%, thou shall not escape the fact that thy model is an approximation."

Why Models Are not Believed: The Croesus Syndrome

Fig. 1.3 The Pythoness of the Oracle of Delphi engaged in her job. A painting by John Collier (1850–1934) presently at the Art Gallery of South Australia

It is said that King Croesus of Lydia, who lived during the 6th century BCE, consulted the oracle of Delphi before embarking in a war on the neighboring Persian empire. The response of the priestess of the oracle, the Pythoness, was, "if you invade Persia, you'll destroy a great empire." Croesus took it as a favorable prophecy and proceeded with his war plans. But the Pythoness had not specified which empire would be destroyed: it turned out that it was Croesus' (Fig. 1.3).

Many ancient stories look a little silly to us, including this one. Couldn't Croesus have been a little more careful? Yes, but the point of these stories is not to be faithful chronicles of real history, they are meant as illustrations of human wisdom or, perhaps more often, of the lack of it. In the case of Croesus's story, probably the idea was to show some common mistakes that people make when planning for the future. When having to choose between different possible outcomes of a certain action, people tend to embrace the most favorable one—it is one of the manifestations of what we call today, "motivated reasoning". We may call "Croesus Syndrome" the tendency of people to be affected by emotional factors that lead them to believe what they like to believe.

It is not difficult to understand what led Croesus to be so badly misled by the prophecy. Place yourself in the King's sandals: when he consulted the oracle of Delphi, he probably had already assembled his troops: lancers, archers, slingers, charioteers, cataphracts, hoplites, peltasts, and whatever kind of people armed with oversize butcher knives or similar things he could pay to fight for him. At that point, imagine that the oracle had told Croesus something like, "look, buster, the Persians will cut your army to pieces and make a well-cooked hamburger out of your royal presence" (which is, by the way, exactly what happened). Could you imagine Croesus coming out of the cleft in the rock of the Pythoness and telling his troops, "sorry, folks, it was all a mistake. Please disband and go back home"? No, the only thing he could tell his followers was, "the oracle told me that God is in our side" or something like that. Propaganda is not just a modern invention.

The story of Croesus and the oracle of Delphi is mainly a legend, but he surely was not the only leader in history who was careless in his military plans. Just think of how Napoleon, and later Adolf Hitler, thought that invading Russia in winter was a good idea. If you want a detailed historical example of poor military planning, you could do worse than reading Mario Cervi's book "*The Hollow Legions*" (1966), the story of how Italy's *Duce*, Benito Mussolini, pushed Italy into the mad enterprise of invading Greece in 1941. We still have the minutes of the meetings of Mussolini's cabinet and we can read how the great leader had made the lethal mistake of surrounding himself with yes-men who vied against each other trying to please the big boss by over-estimating the fighting capability of the Italian troops. On his side, Mussolini was stupid enough to believe in what they were telling him. The result was that the predictive capabilities of the Italian High Command of the time turned out to be no better than those of King Croesus, many centuries before, and only the intervention of the German army forced the Greeks to surrender and saved the Italians from complete humiliation. There are many more cases of incredible strategic mistakes in actions carried out by leaders who, apparently, were swayed by their own propaganda. But this kind of mistakes is very common in all areas of human activity: companies go bust, people lose their money, families break up and more.

We see how emotions play a big role in the way we see the future, and they may lead to disbelieve good models, just as to trust bad ones. We already saw in a previous chapter how people tend to rush to judgment on the basis of the "law of small numbers" described by Twersky and Kahneman [8], showing that emotional factors surely play an important role in leading people to interpret the future on the basis of grossly insufficient data. But they can do worse than that and the same authors, Twersky and Kahneman, explored various forms of misperceptions affecting the human mind [33]. The main ones are: (1) representativeness, (2) availability, and (3) anchoring.

Representativeness is the tendency of people to judge according to stereotypes based on some representative case. For instance, the common image of a university professor is someone with a beard and glasses. Then, when people see someone with a beard and glasses they may assume that he is a university professor. It happens to me all the time: apparently I fit the image of how a university professor should look like. But that is an inference made on the basis of insufficient data: what if I were a drug dealer, instead? Yet, my students never feel that they need to ask my ID when they appear in class for my lecture. Then, *availability* means to judge on the basis of available experience only. For instance, most of us never experienced a tsunami and so we tend to see the frequency of such an event as smaller than it actually is—to say nothing about the possibility of a climate tipping point for the Earth's climate. Finally, *anchoring* means to judge on the basis of available data (the anchor) independently of their significance. A classic manifestation of this mistake is when you have an especially cold day and some people seem to think it disproves the concept of global warming. In January 2019, after an especially harsh cold spell in the Midwest, president Donald Trump tweeted, "What the hell is going on with Global Warming?" Probably, he thought it was funny.

But perhaps the worst and most diffuse reason for the Croesus syndrome is the one called "Groupthink." Not only do people tend to be individually gullible, but they seem to be affected by a dangerous collective phenomenon that makes groups even more gullible than individuals. The term was used for the first time by William H. Whyte Jr. in an article which appeared in *Fortune* magazine, in March 1952. Whyte extrapolated the concept from George Orwell's definition of "doublethink" in his famous novel "*Nineteen Eighty Four*" (1949), but the two concepts are very different. Orwell defined doublethink as the capability of people to hold two different and conflicting views without being able to notice the contradiction (say, you are not racist but you do not want your daughter to marry an Italian man). Groupthink, instead, is the inability to maintain or express one's beliefs in the face of contrasting beliefs held by the majority of the group one belongs to.

Whyte may have been influenced by a series of classic studies on the effect of social pressure carried out by Solomon Asch in the late 1940s and early

1950s. These tests are still known today by the names of the "Asch Paradigm" or the "Asch Experiment. [34]" Their results were always consistent: most people tend to modify their behavior in such a way as to avoid creating disagreement within the group, being singled out as troublemakers, or just making ripples. Not everybody is subject to group pressure in the same way, and Asch found that some people fought back, while others simply withdrew to an agnostic position. But we may imagine that when a group is formed with a certain purpose then, gradually, the outliers will be expelled or marginalized and those who remain will share the basic tenets of the group.

Groupthink may be useful to ensure a certain consistency in the actions of a group and it may be worth nothing that in occultism there exists the term "egregore" indicating a collective manifestation of the thought of the group. Egregore does not necessary have the negative ring of the term "groupthink," on the contrary it may indicate the capability of the group to go beyond the limits of the single persons in it. Nevertheless, groupthink may lead to disasters if most members of the group tend to be optimistic about the outcome of some task. In that case, nobody will want to be the doom-monger who spoils everything. This is another form of the characteristic "feedback effect" of complex systems. The more people are optimistic, the more they tend to infect their colleagues with the same optimism. This is especially true for leaders who tend to control their followers by appearing confident in the success of whatever enterprise they are engaged in.

There is another emotionally based kind of misuse of models that can be seen as the opposite of the Croesus syndrome, the tendency of overestimating the chances that a model may be wrong. It often takes the form of disbelieving negative predictions and we see it at work in particular with climate change. In this field, a common reaction is to scoff at the results of climate models and at the "alarmists" who spread them. It is something we could call "doom fatigue," a syndrome that does not seem to have been quantitatively studied but is well known. An example is the story of the boy who cried wolf. As told by Aesop, the story faults the boy for having called wolf too many times simply because he was bored. But that might just have been bad press and the boy was simply doing his best but fell victim to an especially unlucky streak of events and, more than that, to the tendency of people to disbelieve bad news.

The story of the boy who cried wolf is worth examining in some detail. Suppose that the boy sounds the alarm every time he sees something that looks like a wolf. Of course, the creature might just be a dark sheep or a dog, but the boy does not want to take the risk of missing a real wolf coming. The villagers know the uncertainty in the boy's task and they are prepared for a few false alarms. So, the first time they rush to the village fence and find no wolf there, they pat the boy on the head and say, "keep at your task, boy, we are with you." Then, there comes a second false alarm, the villagers rush to the fence and there is no wolf. "It is fine, boy, but try to be more careful, will ya?" Then, the third night when it happens, "well, boy, this is sort of strange, isn't it? You say there was a wolf? Really? We couldn't see any wolf. You know, you really should be more careful before calling all the villagers to the fence." Then, there comes a fourth night when the rush to the fence goes blank: no wolf. At this point, somebody starts suspecting that the boy is really cheating them. "How do you explain that we ran to the fence four times in a row and the wolf was never there? Boy, either you are making fun of us or you work for the wolf."

But the boy may have been just trying to do his best. An unfavorable streak of bad predictions is not impossible and, in the long run, is unavoidable. Assume that the boy is right most of the times, let us say two times out of three (66%). Not a bad performance but, even so, the probability of a streak of four false alarms in a row is over 1% and it is bound to happen, sooner or later. Consider also that by the time the villagers arrive, the wolf may be gone and what proof is there that there really was one? It does not matter whether the boy sends false alarms or they are just perceived as false alarms. At some point, doom fatigue will set in and the shepherds will probably think that the boy has been cheating them. They will fire him and think that the wolf cannot be such a serious threat, after all. Until the wolf comes for real.

We see this effect at play on multiple occasions when the threat is known but its occurrences are widely spaced in time. Tsunamis, hurricanes, forest fires, are all examples of situations in which the intervals between disasters lull people into a false sense of security. Somehow, the fact that no hurricane hit the city for many years, means that the city is safe from hurricanes and many people will scoff at those doomsters who try to say that precautions should be taken.

"Doom fatigue" is in several ways akin to the phenomenon called "The Gambler's Fallacy" that leads some people to play the Martingale strategy that consists in doubling the bet at every loss, reasoning that if something has not happened up to that moment, then it is more likely that it will happen. Doom fatigue changes sign into this wrong view of probability, assuming that if something bad did not happen so far, then it is more unlikely (or perhaps impossible) that it will ever happen. Translated into the real world debate, you see it at work in plenty of cases.

Consider the data below for US North Atlantic cod fishery [35] (Fig. 1.4).

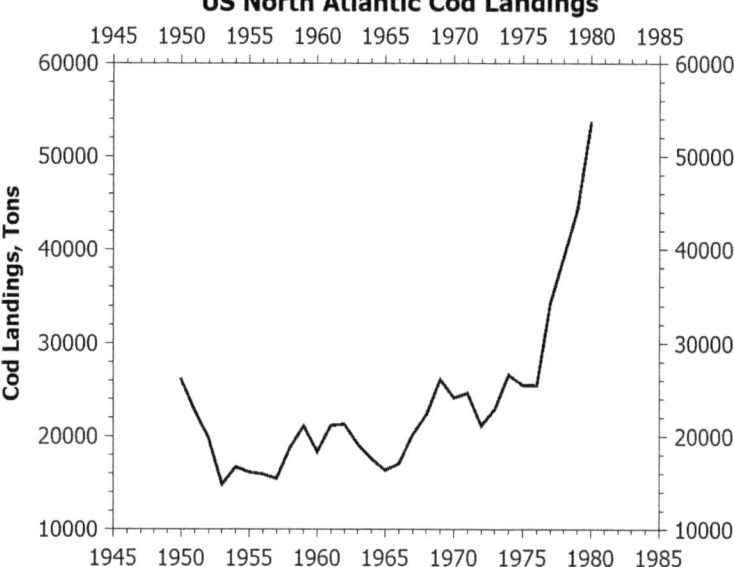

Fig. 1.4 Landings of the US North Atlantic cod fishery. Data from Faostat

Imagine you are the CEO of a cod fishing company: these results would look great to you. Most likely, your company followed the overall trend and managed to double its output in just a few years. That was the result of adopting the new technology of "dragging", allowing fishermen to purse cod in the open ocean and opening up a new resource that was not available to the people practicing inshore fishing. It was a great time for the industry, Hamilton et al. [35] report that during this period, in Canada,

> At the height of the boom, dragger captains made $350,000–600,000 a year from cod alone. Sharemen, many of them high school students on their fathers' boats, could earn $50,000 a year. The federal government helped finance boat improvements, providing grants covering 30–40% of their cost.

But something was deeply wrong in this abundance: the fishing industry was in full overexploitation mode. According to a study performed by Hamilton and coworkers in 2004 [4, 35]:

. . . the main ecological transformation took place during the fishery's high years, not its terminal phase. When some dragger skippers noticed that cod were becoming smaller and harder to find in the mid-80s, they adapted by illegally lining their nets with smaller-size mesh (Palmer & Sinclair, 1997), effectively targeting the juvenile fish. Through this and other intensifications, catches in the final years remained deceptively high despite crashing stocks.

Finally, depletion took its toll and if you look at the cod landing data up to 2012, you can see the Seneca Cliff hitting the US fisheries. Again, the same disaster was befalling Canadian fisheries at the same time, with an even sharper cliff [35] (Fig 1.5).

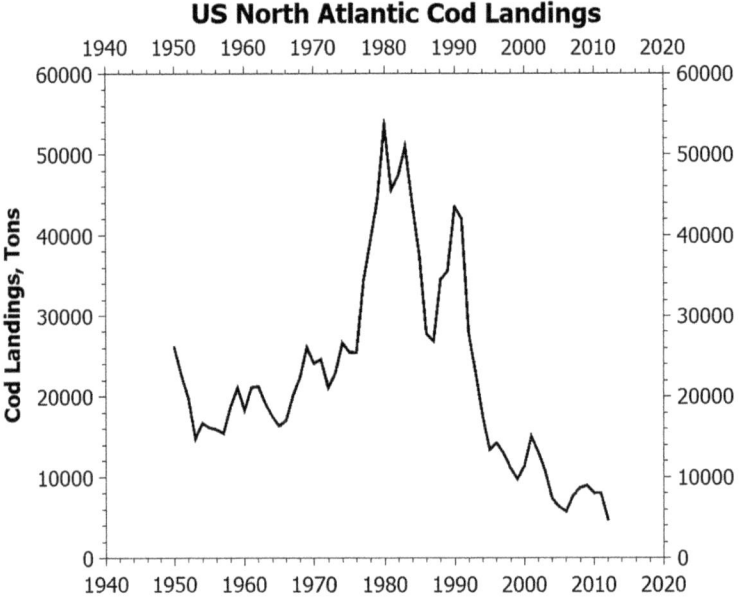

Fig. 1.5   Cod Landings of the North Atlantic Fishery. Data from Faostat

Note that nothing was done to try to avoid the cliff as long as the catches were growing, even though it must have been known to the government that the fishermen were illegally catching juvenile fish. It was only when production started to crash down that the governments intervened with moratoria and quotas. But it was too late: the cod population did not rebound and, today, the North Atlantic fishing industry is mainly catching invertebrates such as crabs and shrimp.

Again, the data could have been telling a lot to the fishing industry if interpreted in terms of a correct model of the system. Knowing that fishing was depleting the cod stocks it was clear that some kind of downturn was to be expected, most likely a Seneca-style cliff, the typical mode of decline of biological populations. The exact start of the cliff was not exactly predictable but there was no doubt that it was looming. Evidently, something went wrong in the perception of the system on the part of the policy-makers, at least in terms of public statements. It seems that individual fishermen were perfectly aware of what was going on, but the deadly mechanism of group-think prevented this knowledge from surfacing and having an effect on policies.

This is a major problem with all models. No matter how good the model is, it is useless if it is not believed. Typically, models telling people that they have to change their ways are the most likely to be disbelieved or ignored. That happens in particular with the models used for climate change, the results either being rejected or accepted but not acted upon. Maybe climate scientists would have more success if they were to wear the robe of a Pythoness and speak from a crack in the rocks, but that has not been tried so far.

References

1. Wack, P.: Scenarios: uncharted waters ahead. Harv. Bus. Rev. https://hbr.org/1985/09/scenarios-uncharted-waters-ahead. Last accessed, Aug 31, 2019 (1985)
2. Ruppert, M.C.: Crossing the Rubicon: the decline of the American empire at the end of the age of oil. New Society Publishers (2004)
3. Musa, G.: Terrestrizzazione. In: Musa, G., Cremaschi, I. (eds.) I Labirinti del Terzo Pianeta. Nuova Accademia Editrice (1964)
4. Forrester, J.W.: Counterintuitive behavior of social systems. Simulation 16, 61–76 (1971)
5. Shields, L.B.E., Hunsaker, J.C., Stewart, D.M.: Russian roulette and risk-taking behavior. Am. J. Forensic Med. Pathol. 29, 32–39 (2008)
6. Ellemberg, J.: How the financial markets fell for the martingale, a 400-year-old sucker bet. Slate. Available at http://www.slate.com/articles/life/do_the_math/2008/10/were_down_700_billion_lets_go_double_or_nothing.html. Accessed 2 Sept 2016 (2008)
7. Phillips, D., Welty, W., Smith, M.: Elevated suicide levels associated with legalized gambling. Suicide Life-Threat. Behav. 27, 373–378 (1997)
8. Tversky, A., Kahneman, D.: Belief in the law of small numbers. Psychol. Bull. 76, 105–110 (1971)

9. Meadows, D.H., Meadows, D.L., Randers, J., Bherens III, W.: The Limits to Growth. Universe Books (1972)
10. Nordhaus, W.D.: World dynamics: measurement without data. Econ. J. **83**, 1156–1183 (1973)
11. Forrester, J.: World Dynamics. Wright-Allen Press (1971)
12. Bardi, U.: The Limits to Growth Revisited. Springer (2011)
13. Nordhaus, W.: Lethal Models. Brookings Pap. Econ. Act. **2**, 1–59 (1992)
14. Solow, R.: Technical change and the aggregate production function. Q. J. Econ. **70**, 65–94 (1956)
15. Turner, G.: A comparison of the limits to growth with 30 years of reality. Glob. Environ. Chang. **18**, 397–411 (2008)
16. Zetetic Method: Rational theoretical standard wiki. Available at https:// rationaltheory.fandom.com/wiki/Zetetic_Method. Accessed 22 Mar 2019
17. Letzer, R.: One conspiracy theory at a time: flat-earthers don't reject climate science. LiveScience. Available at https://www.livescience.com/63470-flat-earth-climate-science.html. Accessed 18 Apr 2019 (2018)
18. Arrhenius, S.: On the influence of carbonic acid in the air upon the temperature of the ground. Philos. Mag. J. Sci. (fifth Ser.) **41**, 237–275 (1896)
19. Bardi, U. Where is the proof that CO_2 warms the Earth? Cassandra's Legacy. Available at https://cassandralegacy.blogspot.com/2018/02/where-is-proof-that-co2-warms-earth.html. Accessed 21 Mar 2019 (2018)
20. Makarieva, A.M., Gorshkov, V.G., Li, B.-L.: Precipitation on land versus distance from the ocean: evidence for a forest pump of atmospheric moisture. Ecol. Complex. **6**, 302–307 (2009)
21. Deshmukh, S., Murthy, J.: Nightfall: Can Kalgash Exist, arXiv:1407.4895 (2014)
22. Laskar, J., Gastineau, M.: Existence of collisional trajectories of Mercury, Mars and Venus with the Earth. Nature **459**, 817–819 (2009)
23. Palyulin, V.V., Chechkin, A.V., Metzler, R.: Levy flights do not always optimize random blind search for sparse targets. Proc. Natl. Acad. Sci. U. S. A. **111**, 2931–2936 (2014)
24. Zagorsky, J.L.: Economics Bulletin, **36**(1), 401–410 (2016)
25. Saltelli, A.: Sensitivity analysis for importance assessment. Risk Anal. **22**, 579–590 (2002)
26. Papert, S.: Mindstorms: Computers, Children, and Powerful Ideas. Basic Books, NY (1980)
27. Taleb, N.: The Black Swan. Random House (2007)
28. Gladwell, M.: The Tipping Point: How Little Things Can Make a Big Difference: Malcolm. Back Bay Books (2002)
29. Sornette, D., Ouillon, G.: Dragon-kings: mechanisms, statistical methods and empirical evidence. Eur. Phys. J. **205**, 1–26 (2012)
30. Hansen, J.: Climate catastrophe. New Sci. **195**, 30–34 (2007)

31. Goldblatt, C., Watson, A.J.: The runaway greenhouse: implications for future climate change, geoengineering and planetary atmospheres. Philos. Trans. A. Math. Phys. Eng. Sci. **370**, 4197–4216 (2012)
32. Antilla, L.: Self-censorship and science: a geographical review of media coverage of climate tipping points. Public Underst. Sci. **19**, 240–256 (2010)
33. Tversky, A., Kahneman, D.: Judgment under uncertainty: heuristics and biases. Science **185**, 1124–1131 (1974)
34. Asch, S.E.: Opinions and social pressure. Sci. Am. **193**, 31–35 (1955)
35. Hamilton, L.C., Haedrich, R.L., Duncan, C.M.: Above and below the water: social/ecological transformation in Northwest Newfoundland. Popul. Environ. **25**, 195–215 (2004)

2

Complex Systems and the Science of Collapse

She crushes the mountain to garbage,
scattering the trash from dawn to dark,
with mighty stones she pelts,
and the mountain,
like a clay pot
crumbles
with her might
she melts the mountain
into a vat of sheepfat
Enheduanna (ca 23rd century BCE) [1].

Complex Systems: The Goddess' Wrath

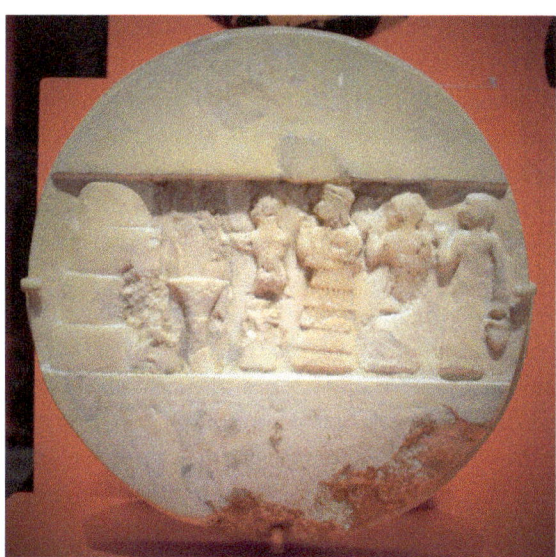

Fig. 2.1 An alabaster disk showing the priestess Enheduanna (third from the right) who lived in Mesopotamia around the 23rd century BCE. (Picture courtesy by Mefman00, https://en.m.wikipedia.org/wiki/File:Enheduanna,_daughter_of_Sargon_of_Akkad.jpg)

© Springer Nature Switzerland AG 2020
U. Bardi, *Before the Collapse*,
https://doi.org/10.1007/978-3-030-29038-2_2

The Sumerian priestess Enheduanna, who lived during the 3rd millennium BCE, is the first known author in history. By some miracle, not only some of her texts have survived to our time, but we even have a portrait of Enheduanna herself in a bas-relief on an alabaster disk recovered from the ruins of the city of Ur. We can see that she a had a strong nose, wore an elaborate garb, and stood in an evident posture of authority as high priestess (Fig. 2.1).

Enheduanna's texts are unlike anything we are used to calling "literature" today and their meaning is not easy for us to understand. But there is one, normally referred to as "*Inanna and Ebih*," [2] that we can recognize as the first report of an ecological catastrophe in history, even though seen with the cultural filters of the time. In the story, we read how the mountain Ebih was a lush and prosperous land, a true paradise of plants and animals. But the Goddess Inanna accused Ebih of "lack of respect" toward her. She donned her weapons, flew into the sky, and smashed Ebih to the ground, "melting the mountain into a vat of sheepfat."

As I argued in a 2015 paper [3], it is not difficult to see in this story a report of the catastrophic runoff of the fertile soil from an overexploited or overgrazed land. It is a process still going on nowadays in the lands once inhabited by the ancient Sumerians, today called "Iraq." In our times we still tend to attribute this kind of disaster to evil entities, including dictators, financial cabals, political conspiracies, religious sects, and more. Of course, it is not possible to prove that evil forces do not exist, just as it is not possible to prove that the Goddess Inanna does not exist. But a probably more rewarding approach to understanding catastrophes is to describe them in terms of the science of *complex systems*, entities formed of subsystems strongly interacting with each other. It is a science that can tell us that ecological catastrophes of the kind that destroyed the mountain called Ebih are the result of the mismanagement of the fertile soil, a fragile entity that is easily washed downhill by rain.

The science of complex system is relatively new because the tools needed to develop it became available only with the second half of the 20th century. In earlier times, physical sciences had been mainly studying those systems that could be described with single equations, for instance the motion of a body in a gravitational field. But that is not possible with complex systems, where each element is coupled to several other elements with ties of comparable intensity. Think of a landslide: each pebble of a pile is kept in place by other pebbles surrounding it. Then, imagine that an external force, maybe a human foot stepping on a pebble, causes it to move downhill. That may free more space for other pebbles to move down and, in turn, each new pebble sliding down may cause more pebbles to start sliding, eventually causing a landslide. It is a phenomenon called "feedback." It is not difficult to understand but, in most cases, describing it with a single equation is simply hopeless.

A landslide is an example of the effects of "enhancing" (or "positive") feedback. It is a reaction of the system that tends to amplify the effects of a perturbation. This kind of feedback can generate spectacular collapses: just as a pebble may start a landslide, a match can trigger an explosion, a shot can start a war, a straw can break a camel's back, and there are many more examples. There also exists another kind of feedback, called "damping" or "negative" feedback, that tends to stabilize the system. Think of a running car: the more you try to accelerate it, the more friction and aerodynamic effects will act in order to slow it down. In general, we can say that feedbacks are the defining elements of a complex system. A system is complex if, and only if, it shows strong feedback effects.

We deal every day with complex systems: animals, people, organizations, and more. It not difficult to understand what is complex and what is not: it depends on whether the reaction to external perturbations is dominated by feedbacks or not. Think of a rock compared to a cat. A rock is not a complex system: kick it and it will just roll forward, the rock has no feedback effects that would enhance the effect of the kick. So, you could write an equation that describes the motion of the rock, taking into account its mass and the friction generated by its movement on the ground. But kick the cat and a variety of reactions could ensue, including a chance that Kitty will turn around and bite you—that is feedback! It goes without saying that there is no "cat equation" describing the behavior of a cat. Indeed, you could say that the best model of a cat is a cat.

Most living creatures are complex systems, including human beings. The same is true for most of the social and economic structures humans create: families, tribes, companies, states, armies, and the like—they are all complex systems too. There are no simple equations describing these systems which tend to behave in ways that can be both unexpected and destructive. In 2003, bringing democracy to Iraq seemed to be easy at the beginning but the system reacted in an unexpected way (although, in hindsight, perhaps not so much).

So, complex systems are restless, never stopping, always changing. They quiver, they vibrate, they oscillate, they crash, and they may collapse and fade away: it is the way they are and they do so because they are *alive*, in the sense that they teem with energy. No complex system can do anything interesting if it has no energy available: a cat without metabolic energy is a dead cat, and a dead cat is not a cat. And that is why complex systems are so interesting, fascinating, even beautiful, although sometimes they are of a kind of beauty that kills (think of a wild tiger or of a hurricane!).

To understand what makes complex systems move and sometimes collapse, we now need to go deeper into the matter and to know a little of the jargon of the field. Below, you will find a list of the most common terms used. None is difficult to understand and you should have at least some idea of their meaning. Here, I am using the term "system" as meaning "complex system".

(1) *State*: the ensemble of the parameters that define the system.
(2) *Forcing*: an external perturbation that tends to change the state of the system.
(3) *Feedback*: the way a complex system reacts to a perturbation. It may amplify the perturbation (positive or enhancing feedback) or dampen it (negative, damping, or stabilizing feedback).
(4) *Attractor*: a set of parameters (a "state") that a system tends to attain.
(5) *Homeostasis (resilience)*: the tendency of a complex system to maintain its state remaining close to an attractor even when it is perturbed by external forcings.
(6) *Carrying capacity*: the maximum flow of energy that the system can maintain for a long time.
(7) *Overshoot*: the tendency of the system to generate flows larger than the carrying capacity.
(8) *Phase transition*: the tendency of a system to jump from one attractor to another, often abruptly.
(9) *Tipping point*: a set of parameters that mark the point that will lead the system to jump from one state to another, switching from one attractor to another.
(10) *Trophic chain*: the tendency of a system to be formed of a chain of linked elements that exchange energy with each other.
(11) *Collapse*: a phase transition that leads to a state of reduced complexity, typically being rapid and abrupt.

Most of these terms are simply related to common sense concepts. We can start with *state*, the ensemble of the parameters of the system. In the case of a living being, the parameters could be the size, the temperature, the metabolic process rates, and more. Then, a *forcing* is any perturbation coming from outside that tends to change the state of the system, it may be physical pushing or pulling, gravitational or electric fields, anything that acts on the system. A complex system reacts to forcings by means of *feedbacks*. As the name says, feedback is the tendency of the elements of a system to influence each other. It means that changes in an element generated by forcing will affect the other elements of the system and not just that, also the element that started the perturbation will be affected back. When enhancing feedbacks dominate the system, it may expand, grow, move, and even explode. When damping feedbacks dominate, it will slow down, shrink, return, and stop, often, the result is that the system will tend to reach a condition of balance near a set of parameters called the *attractor*. It may actually reach those parameters and then it will more or less stay there, or it may oscillate around the attractor the way a pendulum does. This tendency to move toward, and stay close to, an attractor is called "*homeostasis.*" Note that this tendency is not normally the result of a conscious attempt of the system to reach a certain state. It is just the fact that the damping feedbacks tend to maintain it there. Think of a flock of flying birds: there is no "general bird" who gives the order, "now, flock!" It is just that if a bird finds itself outside the flock, it will probably feel uncomfortable enough that it will fly back to the safety of the group. This is a damping feedback that prevents the flock from dispersing into many birds flying independently. But no bird has the flock in mind when they form one.

We also use the term *resilience*, borrowed from engineering, that we can define as the capability of the system to resist collapse. Note that homeostasis does not mean that a complex system is static—it is *not* in "equilibrium" as it is defined in physics and thermodynamics. Equilibrium means that the system has reached a condition of minimum potential energy: if you like, you can say that it has maximized its entropy. A dead cat is in equilibrium, a live cat is in homeostasis. Live cats will normally resist all attempts to skin them, no matter in which way. Maybe you also remember the concept of "strange attractor" mentioned in the first movie of the *Jurassic Park* series. Strange attractors imply that the system is "chaotic" in the sense that it will never return exactly to the same set of parameters it had attained at a certain moment. It is part of the field called "deterministic chaos," a fascinating branch of physics sometimes defined as "Chaos Theory." Here, we do not need to go into the details, it is sufficient to say that many attractors are not strange and that you don't need a system to be chaotic for it to be complex.

Population in real ecosystems is an example of the concept of attractor. There has to be a certain population level which is optimal for a certain species as a function of the available food. It is something we call the "*carrying capacity*" of the system. In biology, it is defined as "the maximum population size that the environment can sustain indefinitely." In homeostatic conditions, the population should never stray away too much from that number: if it becomes higher, food will become scarce and famines or sickness will bring it back to smaller numbers. If it becomes smaller, abundant food will lead it to grow again. So, you see how the complex characteristics of the system create the attractor which, in this case, takes the form of a certain population level.

In practice, in real ecosystems things are much more complicated than just small oscillations near homeostasis. Rather than gently approaching the carrying capacity and staying always close to it, populations may grow so fast that they *overshoot* the sustainability limit of a large factor. Think of a flu epidemic in humans: the flu virus sweeps through the population in periodic waves, its population never stabilizes near an attractor, it rapidly reaches huge numbers by infecting most of the human population, and then it crashes down because it has no more victims to infect. Mother Nature, also known by the name of "Gaia," is far from being a benevolent and merciful goddess. She is rather like the Goddess Inanna as described by the poetess Enheduanna, a ruthless creature who destroys entire mountains. Gaia may do the same when she destroys entire populations that have exceeded the carrying capacity of the ecosystem. It is another characteristic of complex systems: they always kick back and, sometimes, they kick back with a vengeance (cats may also bite you if you kick them).

The oscillations typical of complex systems are bad enough when seen from the viewpoint of the creatures (humankind) who have to go through them. But things may get much worse if the system goes through a *tipping point*. In this case, the system does not return to the original attractor but moves irreversibly away from it to stabilize near a different attractor. The new attractor may correspond to a condition of energy flow and complexity much smaller than that of the previous attractor—in this case, we have a truly disastrous collapse: the Seneca Cliff in its true form.

There are plenty of examples of ecosystems destroyed by human exploitation which did not return to their original state and probably will not do so for a very long time, if ever. For instance, destroying the top predators of the oceanic ecosystem, fish, has created a true explosion of the populations of creatures at the bottom of the trophic chain, jellyfish and crustaceans, once kept in check by fish. Now, these creatures prey on newborn fish, preventing their predators from reappearing. The system is in a different homeostatic condition: one that revolves around an attractor that exists with a much larger population of jellyfish and a smaller one of fish. And so, today, if you are stung by a jellyfish (technically, a *cnidarian*) while you swim in the sea, you know that it is because the marine ecosystem is a complex system!

You see how rich and varied the behavior of complex systems can be. It is a consequence of the fact that they are in a certain way "alive." They have the basic characteristic of living beings, that of being "dissipative structures," a term invented by Ilya Prigogine about half a century ago [4]. Prigogine used the term "dissipative" to evidence that these systems tend to dissipate energy. To be more precise, it is not so much a question of energy but of *energy potentials*, the term used for those forms of energy that can be dissipated. You know what "voltage" is, it is an electric potential. The larger it is, the faster it tends to be dissipated and that may dangerous. If you touch a high potential wire, the potential will be dissipated through your body, another example of how complex systems can kill. But, if it is dissipated in controlled conditions it will power appliances, electronic devices, electric motors, and more.

Electric potentials are just one of the many kinds of energy potentials we can find around us: there are chemical potentials, gravitational potentials, thermodynamic potentials, and more. Sometimes they can be measured using physical units and sometimes qualitatively defined, as when we talk of a person having "sex appeal," referring to her or his reproductive potential. But the existence of energy potentials that can be dissipated is a necessary condition for complex systems. So, a plant is a system that dissipates the energy potential of solar light. A herbivore is a creature that dissipates the chemical potential contained in plants, a carnivore dissipates the energy potential contained in the flesh of a herbivore, a company is a system that dissipates the economic potential of the products it sells, and there are many more examples.

Often, energy potentials are dissipated along a *trophic chain*, a concept that comes from biology but that we can apply to many other kinds of systems. It means that the energy potential is dissipated along a step-by-step chain. You probably know the lines by Jonathan Swift (from *On Poetry: a Rhapsody* (1733)):

So, naturalists observe, a flea
Has smaller fleas that on him prey;
And these have smaller still to bite 'em,
And so proceed ad infinitum,

In biology, the highest tropic chain element (for instance, predators) corresponds to the lowest potential in thermodynamic terms. That may be a little confusing, but it is just a different way of seeing these chains: they are a cascade of dissipation structures. We can also see an industrial system as a trophic chain: it starts with the extractive industry which produces mineral commodities, then these commodities are processed by the manufacturing industry and transformed into products which are finally processed into trash by the waste management industry. Here, too, the main end product is waste heat.

The potentials that sustain the tropic chain are often measured in relative terms according to the concept of EROI or EROEI, the energy returned for energy invested, a term introduced for the first time by Charles Hall in 1988 [5], but also by means of other, similar parameters, such as the "transformity" proposed by Howard Odum at about the same time [6]. The EROI is also related to the concept of "Net Energy," the energy potential that remains available after a transformation. These concepts are normally used to evaluate the efficiency of what we call an "energy production" technology, although we should understand that energy cannot be produced, it can only be transformed. In any case, the EROI of, say, a photovoltaic plant is defined as the energy that the plant produces over its lifetime divided by the energy needed to build, maintain, and eventually dismantle the plant. As should be obvious, when we deal with a commercial enterprise, the EROI should be larger than one, if possible much larger. Nobody would want to build energy plants that produce just enough energy to pay for their construction and maintenance: the plants must produce excess energy that can be sold and provide a profit for the plant owner.

There are many studies in the scientific literature about the EROI of different energy technologies and here it would be out of place to go into the details of the field. It will suffice to say that the current data tell us that renewable technologies such as photovoltaics (PV) now have an EROI comparable to, or even larger than, that of fossil fuels [7, 8]. But the EROI is much more than a commercial parameter. It is a fundamental concept in biology as a measurement of how the dissipative structures of a complex system exchange energy with each other. If a transformation has an EROI larger than one, it means that the trophic level involved grows (it absorbs more energy than it dissipates). If, on the contrary, it is lower than one, the trophic level shrinks. That can be explained with a biological example: you can see a lion as a machine that transforms gazelles into more lions by eating them and later reproducing. Now, a lion gathers energy by eating gazelles but must spend energy to chase them. If the energy the lion spends is larger than what it obtains, then the beast has EROI smaller than one and it must die of starvation and the size of the pride will shrink. If it is larger than one, it has a chance to reproduce and increase the number of lions of the pride. Overall, the EROI of the transformations involved in the system should remain close to one, even though it is likely to oscillate.

The line connecting EROI and trophic chains is just entropy doing its job of increasing all the time as it should do according to the second principle of thermodynamics. That is the way the universe moves, the basic and the only mechanism that makes things move: potentials accumulate and then dissipate, sometimes very rapidly. It is just another way to say that nothing would ever fall down if it had not climbed up before. The fact that the fall tends to be fast is a manifestation of the principle called "the maximum entropy production" (MEP) principle [9]. The production of entropy can be seen as another way of stating that energy potentials are dissipated.

The MEP principle is fundamental for what we are discussing here: it tells us not only that energy tends to be dissipated, but that it tends to be dissipated *fast*, and that is the origin of all collapses. That is obvious when you think of mountain climbing, rope walking, or similar kinds of activities involving the risk of falling down from the height of a gravitational potential. It is a little less obvious when you deal with systems which, normally, are not supposed to collapse, say, governments and large companies. But there is a deep similarity between climbing a mountain and growing an empire: in both cases, a large "energy potential" is accumulated. When you climb a mountain, you accumulate gravitational energy, when an empire arises it accumulates economic energy in its governing structures. All these systems are subjected to the MEP principle and they may go through rapid phases of energy dissipation, an event that we call collapse. These are the basic thermodynamics principles explaining why "ruin is rapid", as Seneca noted long ago.

The MEP principle is a rather abstract concept that does not tell us the actual mechanism of collapse; but it is the result of the networked structure of the system. As we shall see more in detail in the next chapter, the elements of a complex system are linked to each other and they "talk" to each other. In a collapse, each element that starts moving in a certain direction takes other elements with it and the result is normally a cascade of effects all going in the same direction. It is what I call the "Seneca Crunch," also known to William Shakespeare when he said that "when sorrows come, they come not single spies, but in battalions." that is, the collapsing elements of the system seem to be ganging up together, apparently with that exact purpose. But there is no intelligent purpose in complex systems: it is just the way nature works. In the simplest version of the Seneca crunch concept, all you need is a trophic chain of three elements interacting with each other. The central system of the chain may collapse very fast because of the action of the other two, as I describe in my previous book using the formalism of system dynamics [10].

This said, complex systems can take on a bewildering variety of structures: a trophic chain may be composed just of a few interacting elements: think of a market where a single product is sold in a monopoly regime. Or, it can be composed of thousands of elements: think of a forest ecosystem. In all cases, the feedbacks involved in the interactions among the elements may gang up to push the system to heaven or to pull it down to hell. We shall see this behavior in detail in the coming chapters. For the time being, just remember that feedback-generated change is the typical characteristic of complex systems and that this change may lead the system to explode, shatter, crumble, collapse, flounder, crack, evaporate and more. It is the way the Universe works: collapse is not a bug, it is a feature.

The Power of Networks: The Ghost in the Shell

Fig. 2.2 Ugo Bardi's students at the University of Florence engaged in studying the behavior of complex networks in the form of a hourglass in 2019. (Photo by the Author)

In 1803, the Italian scientist Giovanni Aldini attempted for the first time in history to revive a dead person using electricity [11]. Aldini was the nephew of Luigi Galvani (1737–1798) known, among other things, for having noted how the muscles of a dead frog could be made to contract by means of an electric current. Aldini took his uncle's idea several steps forward, trying it on the fresh cadaver of a hanged criminal. Needless to say, the dead man did not come back to life, apart from some gruesome contraction of his muscles when Aldini passed current through his body. The only long term result of this attempt may have been to inspire Mary Shelley for her novel, *Frankenstein* (1818) (Fig. 2.2).

Aldini's attempt, just as Shelley's novel, can be understood as part of the general cultural view of their times, when modern biology had started to grapple with a fundamental question: what makes living creatures alive? In ancient times, it was commonplace to think that living beings had something inside, an inner force, a "ghost in the shell" that made them move, change, grow, and do all the things that the living do and that the dead do not do, except in zombie movies.

The idea probably derived from the observation that living beings breathe and that when they cease to breathe they also cease being alive. So, the movement of the lungs was seen as generated by the internal pressure of something called *anima* in Latin, a term connected with the Greek one of *ànemos*, meaning 'wind,' and which today we translate in English as *soul*. The Latin term, *anima*, was then incorporated into the concept of "animal," literally something that has a soul, even though nowadays we tend to use the term "soul" for a more abstract, religiously-oriented interpretation.

The idea that living beings have a soul is called "vitalism." It ran into troubles when biologists started dissecting animal and human bodies, an activity that became popular in Europe with the Renaissance. Despite all efforts, nothing could be found inside that looked like a soul. Many people tried to identify the soul with one or another human organ, the hypothalamus and the pituitary gland being especially popular, but that led nowhere as it was gradually discovered that these glands had specific purposes unrelated to the concept of soul. Something else was making living beings alive, but what? It was not an easy question to answer and vitalism remained popular throughout the 19th century, and even later. It is still popular today, even though largely abandoned in scientific circles.

The way we see this issue, nowadays, is related to what we saw in the previous section: living beings are *complex systems* and they show all the non-linear characteristics associated to complex systems. Then, the fact of being alive is an "emergent property" of a complex system. It means that there is no such thing as a ghost in the shell: it is the whole that gives the system the properties that make it complex. There is more: this characteristic is typical of the systems we call "networks," only networks can be complex systems.

Let me make an example: imagine yourself as an ant. You live in an anthill, a network of individual ants, each one interacting with other ants, one at a time. As an ant, you are nearly blind but you have an excellent sense of smell and most of your sensor inputs are *pheromones*, signaling molecules that you receive from your sister ants that then you re-transmit to them.

The exchange of pheromones leads ant colonies to direct their members in various directions for foraging. Ants also show a variety of behaviors: they may attack other colonies, swarm away from their original site, reproduce by creating other colonies, and behave in other fascinating ways. Yet, the colony has no structure that we could see as taking decisions, a brain of some kind (to say nothing about having a soul). In fiction, ant queens and soldiers may be humanized as you can see in movies such as *Antz*, (1998) but, in the real world, neither queens nor soldiers rule the ant colony. The complex behavior of the anthill is the result of the collective interactions of ants with each other. All the worker ants are the same and no ant has the anthill in mind, no ant even perceives the existence of the anthill. The colony is an *emergent property* of a network of ants.

An emergent property is not something material, it is something that we perceive as a pattern. Try a simple test: use a pencil to draw a square on a piece of paper. Now, look at it. What you see are dark graphite particles on the white surface of the sheet. None of these little chunks of graphite is a square or has anything to do with a square. The square is an emerging property of the ensemble of these graphite particles.

Societies formed of human beings tend to generate emerging patterns of all kinds, but single human beings may not be able to understand them exactly. If I were to ask you about the equivalent of the anthill for humans, your country, you would be able to describe no more than a few features of it. Italy, for example, is a mountainous peninsula in the middle of the Mediterranean Sea which, due to some quirks of geological history, looks like a boot. It has some sixty million inhabitants, they speak a language called "Italian," but they seem to find it insufficient to convey what they want to say since they tend to gesticulate a lot. And they are fond of eating cereals in the form of a round loaf called "pizza." And on you could go with many weird things Italians tend to do, but all I could tell you about Italy would always be just a very minor reflection of the incredible complexity of an entity formed of tens of millions of human beings.

In practice, apart from the data you learned at school, your perception of the country you live in is largely shaped by the pairwise contacts you have with other human beings: in this sense, you are not so different from an ant. For humans, these contacts involve mainly verbal and visual signals, not olfactory ones as for ants, but the mechanism of transmission and re-transmission is the same: you are continuously exposed to signals from your fellow human beings, in person or by means of technologies such as the media or social networks and you continuously elaborate and re-transmit these signals to other humans. Outside this realm of pairwise interactions, most humans probably see their city, their country, or the state of which they are citizens mainly as patterns created by simple messages related to defense and attack. We may say, "right or wrong, it is my country," maybe we can get excited listening to the national anthem while the flag flaps in the wind, and, in some occasions, we may feel that it is our duty to charge armed with a bayonet against a machine gun nest. All that is done in the name of a nebulous entity called the "country" formed of a very large number of entities, individuals, families, associations, companies, cities, buildings and much more.

The behavior of the immense network defined as a country is generated by transient bursts of reinforcing signals that may generate rapid, even violent, collective reactions on the part of the whole human colony. To be sure, unlike the ant colony, the human colony (also called "state") does have a brain, of a sort. It is called "government" or, sometimes, "our beloved leader." But it is doubtful that this entity can do much toward steering the human colony toward an intelligent behavior. States explore their environment, compete for resources, occasionally fight each other, at times very destructively. But these are behaviors that ant colonies engage in as well.

So, human societies, ant colonies, and other systems formed of many linked entities are best understood if they are seen as *networks*. The science of networks, just like the science of complex systems, has made large advances in recent times and the two concepts are strictly related: no system can be complex if it is not a network and many—although not all—networks are complex systems.

Before going more in depth into the matter, we need to learn a little of the jargon of the field of network science. Here is a list of the most important terms used.

- *Graph*: a set of objects in which some or all pairs of objects are connected to each other.
- *Network*: a dynamic graph whose elements exchange energy, matter, or information.
- *Node* (also *vertex* or *point*): one of the objects forming a graph.
- *Link* (also *edge*): the connection between two nodes.
- *Size*: the number of nodes in a network.
- *Density*: the ratio of the number of actual links to the total possible links.
- *Degree*: the number of links to a node, defined also as an average among all nodes.

- *Clustering coefficient*: the ratio of actual links connecting a node's neighbors to each other to the maximum possible number of such links. Also defined as an average among all nodes.
- *Path length*: the smallest number of steps necessary to connect a node to another node. Also defined as the average path length characteristic of the whole network.
- *Network topology*: the way the nodes are connected to each other.

As you can see, the field of network science is rich in concepts and the terms listed above are just some of the many parameters that can be defined in network science. As an exercise, let us apply these concepts to an ant colony. In this case, a *graph* would be a snapshot of the anthill, showing an image of all the ants and the underground burrows. The *nodes* of the anthill are, obviously, the single ants. The *links* are the connections that the ants make with each other when exchanging pheromone signals. The whole of these elements forms the anthill network, the *size* of which is equal to the total number of ants. At any given moment, only some ants are connected to other ants, so the density of the network is rather low, just as the clustering coefficient: ants seem to contact each other only pairwise. But ants move and, in principle, every ant can connect to any other ant, so the topology of this network is said to be "*fully connected*." Other networks have immobile nodes and may not be fully connected. In a human brain, for instance, every neuron is connected to a limited number of nearby neurons and this kind of network should be termed a "*lattice*." A signal from any neuron can reach any other neuron in the brain, but not directly, it needs to jump from neuron to neuron.

Lattice and fully connected networks are part of a remarkable fauna of topological arrangements that can be summarized as shown here—an incomplete list!

- *Connected network*: every node of the network is connected to every other node by means of a route of links.
- *Lattice network*: nodes are connected only to their near neighbors in space.
- *Random network (also known as Erdös-Rényi network)*: each node is connected at random with other nodes.
- *Small world network*: nodes are mostly connected to nearby nodes, but also to some remote ones.
- *Scale-free network*: similar to the small world network, but the probability of a connection can be described in terms of a "power law."
- *Fully-connected Network*: every node is directly connected to every other node.

- *Cellular network*: the nodes are arranged in "cells" where the connection between nodes are most frequent, connections between different cells are rarer.

As a last general observation, note that there exist virtual and physical networks. A virtual network exists in virtual space with the nodes exchanging information with each other, typically the network is generated by a digital computer. A real network is an actual assembly of physical objects that exchange energy with each other—we may say that they exchange information as well but, in any case, transporting information from a node to another requires energy. An example of coupled virtual and real networks is that of the World Wide Web and the Internet. These two terms are sometimes used interchangeably, but they are not the same thing.

The interesting point, here, is how networks can generate complex behavior. It can take the shape of the rearrangement of the connections, a variation of their number, an expansion of the network size, and more. Networks can also collapse and, in this case, they may rapidly lose a number of connections or of nodes, or both. Collapse may mean that part of the network "evaporates" leaving a smaller network behind, or the network may break down in two or more smaller networks not connected to each other.

As you may imagine, there is a wide variety of phenomena leading to this behavior and not all networks are complex systems. Some networks are simple mechanical arrangements of elements and they show a linear behavior. Think of a mechanical clock: it is a network in the sense that it is a system of gears (nodes) linked to each other that transfer energy from a spring (or a weight) all the way to the needles of the display. Surely, a clock has something related to the concept of complex system as we defined it in the previous chapter: it dissipates an energy potential, the one stored in the spring. But that doesn't make it a complex system. The same is true for the many mechanisms and robots used in manufacturing and process control: they are designed to operate as linear systems. A mechanical arm working at an assembly chain must—and normally does—behave in a linear way, predictably and reliably: no tipping points, no phase transitions, nothing like the varied and unexpected behavior of a complex system. That does not mean these robots are always harmless: some kill people by mistake such as in the case of Robert Williams, possibly the first human being in history killed by a robot, a mechanical arm in the factory where he was working. But even military robots, possibly the most sophisticated kind existing today, are supposed to be tools rather than independent entities.

In practice, we do not need great mechanical sophistication to create a network with a behavior typical of complex systems. As an example, let us think of a very simple device: the sandglass (or hourglass). Before the age of mechanical clocks, people had several ways to measure time. One of the earliest devices to do that was the water clock, called also *clepsydra*, a poetic name meaning "thief of water" in Greek. In the late Middle Ages, the sandglass was developed, more accurate and easier to manage than the clepsydra. Sand grains behave like cars at a highway gate or people at the checkout of a supermarket: no matter how long the queue is, the throughput rate does not change [12].

As time-measuring machines, hourglasses are linear devices. But there is a non-linear characteristic of sandglasses that nobody had noticed or, at least, written about until recent times: the avalanches in the bottom reservoir. If you ever watched a sandglass in action, it is likely that you found yourself fascinated, or even hypnotized, by how the falling sand forms a pile that grows, reaches a certain height, and then collapses in a small avalanche. The process repeats itself as long as there is sand flowing down from the upper reservoir. If you spend some time watching, you will note that the avalanches may be large or small, and they seem to take place in a random manner.

Nobody had studied the statistical properties of the avalanches in a sandpile until the 1980s, when the Danish physicist Per Bak and his colleagues, two postdoctoral researchers, Chao Tang and Kurt Wiesenfeld, were working at the Brookhaven national laboratory. They had been developing a model of a system of coupled oscillators, something that could have been relevant for solid state physics. But, at some moment, they understood that their model could describe the avalanches in a sandpile. It was a smart marketing idea: the model was the same, but most people can understand avalanches much better than solid state physics.

Bak, Tang and Wiesenfeld published their ideas in 1988 [13], proposing the concept they called "self-organized criticality" (SOC). It was the birth of the "sandpile model," or "BTW" model, from the initials of the authors. It became well known, and was popularized by Al Gore in his book *Earth in the Balance* (1992) and by the fascinating book by Per Bak himself, ambitiously titled *How Nature Works* [14] (1996). The interest of this model is not so much related to the way it could describe sandpiles: real sandpiles do not always generate avalanches with the frequency that the BTW model predicts. The interest of the model was that it could be applied to a variety of phenomena, including earthquakes and wars. Scientists found the concept of self-organized criticality fascinating: it is one of the characteristics of living beings to live always "on the balance," always sliding down an avalanche but never being overcome by it—like surfers riding a wave. That such a simple model could catch this characteristic was remarkable.

So, how does the BTW model work? It assumes that the sandpile is a network that can be represented as a chessboard. Each square of the chessboard is a node and it is linked to four near-neighbors. In the model, each square is assumed to be gradually filled with an increasing number of virtual grains of sand "falling from the sky" on random squares in the grid. The maximum number of grains that each square can hold is fixed—let's say it is equal to three. When a square receives a fourth grain, it "collapses", that is it spills the four grains it contains to the four neighboring squares. Then, one or more of these squares may collapse in turn because they already contained three grains. The excess grains spill to neighboring squares and the result may be an avalanche that spreads over the 2D network until all the excess virtual grains either come to rest in some not critically full squares, or "fall off" the edge of the chessboard grid. When the model is run, the number of avalanches and their size (the number of grains involved) can be recorded.

Now comes the interesting point: the size of the avalanches generated by the BTW model looks random at first sight but it is not. It turns out that large avalanches are less frequent than small ones and that there exists a well-defined relationship—called a "power law"—between the size and the frequency of avalanches. The size is proportional to the frequency raised to a variable exponent—that can also be expressed as "raised to the power of an exponent," hence the name, power law. It means that large avalanches are rarer than small ones but the probability of an avalanche of a certain size to occur depends on the exponent in the equation.

The power law is also called the "fat tail" law to indicate that extreme events (the tail of the distribution) are not so rare as in other statistical distributions, such as the common Gaussian curve. These relations are also called "Pareto's law", from the 19th century Swiss economist Vilfredo Pareto who had proposed it for the first time in 1895 for the distribution of income in Switzerland [15]. You probably heard of Pareto's law in the simplified form often called the "80/20" law, often referred to practical cases: you could hear that in a company 80% of the work is done by 20% of the employees. Or that, in war, 80% of the fighting is done by 20% of the soldiers. Pareto himself proposed this 80/20 relationship when examining the distribution of land among landowners in Italy, saying that 80% of the land is owned by 20% of the owners. But that does not mean that the 80/20 rule is an exact law, it is an approximation that gives some idea of this kind of phenomena. For instance, Jordi Prats reported in the blog of the American Statistical Association [16] that

In the US book business, instead of an 80/20 rule, we find a 97/20 rule, that is, 97% of sales are made by 20% of authors. US literary nonfiction sales are still more imbalanced; with 0.25% of books representing 50% of sales. In Canada, a 0.8% of books generated 60% of bookshop revenues.

These data could tell us something about how many copies a book dealing with collapse could sell, and that might dishearten the author a bit. But it could be worse: there is another version of the law, sometimes called "Sturgeon's law" from the name of the science fiction writer Theodore Sturgeon (1918–1985). Sturgeon's law is expressed sometimes as "99% of everything is crud" or, in its strong form, "99.9% of everything is crud." It is the same idea as Pareto's law, just a bit more extreme. Apart from the numbers, anyway, these laws tell us that large events are much less probable than small ones.

It turns out that the universe likes power laws and many common phenomena can be explained by the concept of self-organized criticality. A characteristic of the BTW model and of other similar models is that the power law is not written anywhere in the program that manages the model, and it does not seem to be possible to derive the power law from the structure of the model except by means of running the program and recording the results. In other words, despite the fact that the programmers know everything about the system they have created, they cannot derive the existence of the power law from the lines of code they wrote. In this sense, we can say that the power law is an "emergent" property of the system. That is, it is a collective characteristic of the system that does not reside anywhere in the properties of single nodes—just like the anthill is not encoded in the brain of single ants.

All this has an immediate relevance when we think of it in terms of collapses. Imagine that the virtual sandpile of the BTW model is inhabited by virtual creatures who are periodically wiped out by the avalanches. They would try to learn what they can about these disasters hitting them apparently at random and, soon, they would learn that there exists a power law linking the probability of occurrence of avalanches to their size. They would also learn that the areas of the sandpile where there are several squares already with 2 or 3 grains are dangerous. They would not be able to predict where the next avalanche would take place nor its exact size, but the model would give them some idea of what to expect and where that would make them able to take precautions. They would also understand that if any local prophet were to tell them that he knows exactly where the next avalanche will be, he should be regarded as a dangerous charlatan. Indeed, self-organized criticality is the way of understanding something that had been known for some time: some phenomena appear to be random but they really are not—not completely, at least.

This is not just an abstract matter: it has great relevance for the real world. For instance, it had been known for a long time that large earthquakes are rarer than small ones, but it was only in 1954 that Charles Francis Richter and Beno Gutenberg [17] noted that there exists a mathematical relation, an "inverse power law," that links the number of quakes to their size. It means that earthquakes are emergent phenomena that follow laws similar to those of the sandpile described by the BTW model. Most earthquakes are small enough to be harmless: if you live in Japan you are likely to experience at least a few earthquakes every year which are large enough to be noticeable, with the furniture around you moving and shaking and the whole building generating ominous noises. It may be a little scary but, normally, nothing happens. After a while, the quaking stops and people go back to their chores. But the "fat tail" law implies that large earthquakes are possible and, indeed, they do occur.

But why do earthquakes follow the Pareto law? Although the law can be expressed with a very simple equation, there is no way to deduct it from first principles, that is from a knowledge of, say, the physics of earthquakes or the structure of Earth's tectonic plates. Again, we see that these laws are emergent phenomena of complex systems. It seems that the universe just likes power laws and networks are common in the universe, too.

Seeing complex systems in the form of networks tells us that collapses are always collective phenomena, meaning that they can only occur in networks of elements connected with each other. So, the things that collapse; everyday objects, towers, planes, ecosystems, companies, empires, or whatever you have, are always networks. Sometimes the nodes are atoms and the links are chemical bonds; that is the case of solid materials. Sometimes the nodes are physical links between elements of artificial structures, that is one of the subjects of study of engineering. And sometimes the nodes are human beings or social groups and the links to be found on the Web or in person-to-person communication, or maybe in terms of monetary exchanges. This is the field of study of social sciences, economics, and history.

At this point, we may consider how we can describe collapse in terms of network science. In the previous chapter we saw how collapse can be seen as a manifestation of the MEP (maximum entropy production) principle that sees complex systems rearranging in such a way to generate as much entropy as possible at the fastest possible speed. How about networks? For sure, the laws of thermodynamics remain valid for them just as for all systems in the universe. We can always see a network as a system where energy moves from one node to another and is eventually dissipated by running away from the system. Collapse in a network can take the shape of a rearrangement of the links between nodes, or some nodes may disappear, or the whole network, once connected, may break into two or more smaller chunks. If that happens in a lattice network—that is in a solid piece of material—we call this phenomenon "fracture."

There are various kinds of rapid rearrangements in networks that can be defined as "phase transitions." In the physical world, phase transitions can take place in solid materials when the bonds between atoms are rearranged to form new and different crystalline structures. That's the case of the "martensitic transition" that transform the relatively soft iron into the hard material we call steel. The blacksmiths of ancient times who were engaged in making good swords did not probably know they were changing the network structure of the metal they were hammering, but that is what they were doing. Then, there are all sorts of phase transitions that can take place in virtual networks, as described, for instance, by Barabasi [18]. Only a few of these transitions can be seen as actual collapses, but it may be worth our while to spend some time describing the mechanism of the kind of network collapses which we define as *fractures*.

The fracture of solids is a complex phenomenon that engineers completely understood only in relatively recent times, mainly with the work by Alan Arnold Griffith (1893–1963) who, in turn, built his theory on the basis of an earlier work by Charles E. Inglis (1875–1952). The basic idea is that fracture is the result of localized stress that develops at those places in the solid where *cracks* exist. Just as the name says, cracks are fissures that separate the atomic network of the solid into two portions not linked to each other. Normally, they are supposed to be small and their effect negligible, but cracks exist in all solids except in very special conditions.

Inglis was the first to study the properties of cracks, and he studied what he called "stress concentration" at the tip of the crack. His mathematical model says that the stress is larger, the sharper the crack is, and that at the tip of the crack it may reach orders of magnitude larger than the average stress in the solid. At this point, it was left to Griffith to calculate how long the crack should be in order for the stress at the tip to be larger than the mechanical resistance of the solid. At that point, the chemical bonds at the crack tip would be broken and the fracture would propagate by the usual MEP mechanism. That makes the solid dissipate the elastic energy it had accumulated in its chemical bonds so fast that "explosive" is a term that describes the resulting collapse.

It turns out that fracture is a typical feedback-generated phenomenon: once a chemical bond at the tip is broken, then the edge of the tip moves to the next bond. That breaks down, too, then the tip moves onward. It is as if the atoms in the network of the solid were telling each other "I can't hold it anymore, now it's your turn." In this way, the crack propagates, eventually causing the fracture. The longer the crack, the more energy is released: the end result is the Seneca cliff of the solid. The critical point that starts the avalanche is called the "critical crack length" or, sometimes, "Griffith crack length." Now you know why a balloon pops when you puncture it with a pin! The small hole you create on its surface is a crack and if the balloon is stressed enough that its diameter is larger than the critical crack length, then—bang!

Griffith's theory was developed for real networks of atoms but it may help us understand the general phenomenon of the collapse of networks, real or virtual. It illustrates some of the elements that are common to all complex systems: non-linear phenomena, maximum entropy production, networking, and that typical characteristic that makes complex systems always ready to surprise us: their unpredictable and sudden switch from one state to another. Within some limits, these characteristics can also be observed in apparently unrelated systems: social, economic, and biological systems, as they are all subject to the general thermodynamic laws that govern the universe. Of course, human beings in a socioeconomic system are not the same thing as atoms in a crystalline solid, but there seems to exist a certain degree of unity in the way the universe works, always ruled by the iron laws of thermodynamics.

Living and Dying in a Complex Universe. The Story of Amelia the Amoeba

Fig. 2.3 The author giving a lecture on the growth and collapse of complex systems. On the whiteboard, you can see a drawing of his pet, Amelia the Amoeba. (Photo courtesy, Sara Falsini)

When I teach system dynamics to my students, I often use a simple example to explain the basic elements of the science of complex systems. It is that of "Amelia the Amoeba," a sort of comic book character. Amoebas are unicellular creatures and, from our viewpoint of multicellular organisms, we may tend to consider them as very simple, asexuated lifeforms. That may not be correct: some amoebas are known to have three sexes [19] and others to engage in complex activities such as swarming and even agriculture [20]. There is one species, the *Naegleria Fowleri*, critters normally feeding on bacteria, that can eat your brain if even a single one is inhaled through the nose. But, here, Amelia and her sisters are just supposed to be an example of a relatively simple ecosystem behaving like many other single-celled creatures in terms of reproduction, that is splitting into two copies. Just for fun, let us assume that Amelia is female (Figs. 2.3 and 2.4).

Fig. 2.4 Amelia the Amoeba. Drawing by Ugo Bardi

Amelia starts her existence, alone, in a small glass container (a Petri dish). She is a predatory creature: amoebas eat smaller organisms or they may be "detritivores", eating dead material. So, let us assume that the container is filled with a nutrient solution. Amelia eats, grows, and, when large enough, she splits into two copies of herself. Then, these two daughters (or maybe sisters) of the first Amelia, split again, this time into a total of four copies. And that goes on again, and again. But, of course, Amelia's daughters can continue growing only as long as there is food available, and that is a problem: a finite Petri dish can only contain a finite amount of food. At some point, the destiny of the poor critters may not be pleasant: they will have to starve or suffocate in their own excreta—possibly both things at once.

Amelia and her children can be seen as a simplified metaphor for the way all complex systems behave: they consume resources, they grow, they may collapse. The mechanisms of growth of a biological population can be described according to models developed over the history of the studies in biology and in economics. I gave recognizable names to these different modes, mainly taken from the people who proposed them or, at least, who were involved with them. So we shall be talking about the "growth modes" of complex systems.

1. *The Solow mode*: Amelias grow exponentially forever.
2. *The Malthus mode*: Amelias grow until they reach a stable population and then stay there.
3. *The Hubbert mode*: Amelias' numbers grow and then decline.
4. *The Seneca mode*: the Amelia population grows slowly then collapses rapidly.
5. *The Hokusai mode*: the collapse of the Amelia population is caused by an external perturbation.
6. *The Lotka-Volterra mode* (also, the "*Seneca Rebound*" mode): the system recovers from collapse and restarts growing in a series of cycles.

Growing Forever: The Solow Mode

Let us assume that there is abundant food in the Petri dish where Amelia lives. The little critter will happily grow and then split into two copies of herself. We start with one amoeba, then there will be 2, then 4, 8, 16, 32 and so on in what we call a "geometric progression," with the number of amoebas doubling at constant intervals of time. This kind of growth in discrete steps is not exactly the same as the continuous kind, called "exponential," but for our purpose we can assume it is. It can be represented in a graph as an upward sloping curve shooting toward infinity. Obviously, this kind of growth involves no collapse. (Fig. 2.5).

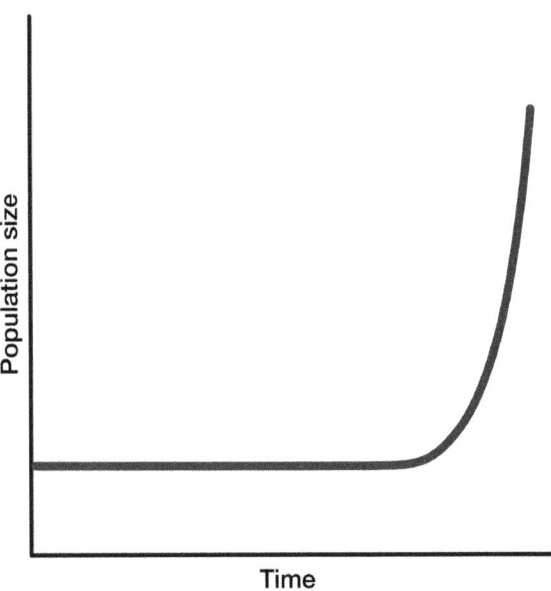

Fig. 2.5 The shape of exponential growth

This exponential growth mechanism of population was already known during the 18th century and was one of the factors that led Malthus to express his theories about the limits to the growth of human population. In economic systems, exponential growth has also been known as the result of constant interest rates. That is supposed to be a good thing for investors, just as for a whole economy. In the 1950s, the American economist Robert Solow was among the first to study the exponential growth of economies [21] using a model called today the Solow-Swan model. This model has been tremendously influential in economic thought and is still popular nowadays. So, we may associate this mode to Solow's name.

There are many examples of exponential growth: as a biological phenomenon, it is typical of creatures such as bacteria and other unicellular organisms. These small creatures can grow very fast and, just for fun, we could calculate how fast the daughters of Amelia could grow in the presence of abundant food. As an approximation, we can say that the generation time for amoebas is 30 min, similar to that of other unicellular creatures [22]. A single amoeba may weigh about 10^{-10} g (that is one-tenth of a billionth of a gram, or 0.1 picograms) [23]. So, if we start with a single amoeba and let it divide undisturbed, the mass of the colony could become one gram in about 33 generations, that is in about 15 h. In 7 more cycles, about two hours more, the amoeba colony would reach the mass of a human experimenter (ca. 100 kg).

Evidently, such a mass could not possibly fit into a Petri dish, and the result would be something like the 1958 science fiction movie *The Blob*, where an initially small, formless creature grows amoeba-like by eating people in great numbers. Of course, that could never happen in the real world, but something similar could take place if a single amoeba of the *Naegleria fowleri* species finds itself inside a yummy human brain. Its descendants would completely eat the brain in a few days. Fortunately, it is a rare disease.

This simple calculation illustrates the power of exponential growth. Biological creatures tend to grow as fast as they can and they are only stopped by limitations in the availability of food and other resources. Unicellular and multicellular creatures behave in the same way, but large creatures, vertebrates for instance, growth at lower rates and rarely exponentially. Ecosystems are nothing like the pristine environment of a Petri dish and, in normal conditions, the growth of biological species is constrained by limited food, predation, and other factors. Nevertheless, there are a few historical examples of an explosive growth of vertebrates. One is that of rabbits in Australia, an infestation believed to have started in 1859 when an Australian farmer named Thomas Austin imported 24 wild English rabbits and set them free on his land. Apparently, he did that just for the pleasure of hunting them and he probably could not even imagine the disaster that he was causing. Within a few years, the 24 rabbits were expanding at a rate of some 70 miles per year, wreaking havoc with the local fauna. By the 1930s, some estimates spoke of 10 billion feral rabbits living in Australia. It is probably an exaggerated number, but it is true that by now rabbits are entrenched in some regions of Australia and are resisting the various attempts of the government to eradicate them by hunting, poisoning, or disease spreading. So far, there has been no Seneca cliff for the Australian rabbits.

Another species that spread exponentially in Australia and created immense damage to the local fauna is the domestic cat. Cats were introduced in the 1800s and now there may exist more than 6 million feral cats all over Australia. So many that they have become food for some Australian Aboriginal communities. Then, of course, there is another invasive species in Australia, the *homo sapiens*, now some 25 million individuals, also responsible for the destruction of much of the original Australian environment. As we all know, humans have been growing exponentially all over the Earth, reaching a population of more than 7 billion individuals, nowadays. It is, fortunately, tapering off but it is not obvious that the slowdown will save this species from an upcoming Seneca cliff caused by overpopulation.

Examples of exponential growth exist also outside biology. One from physics is the chain reaction that leads to nuclear explosions, a fitting example for a book dealing with collapses. These nuclear reactions occur because the atomic nuclei of some elements, in particular uranium and plutonium, may exist in a "fissile" form. When they are hit by a low-energy neutron, the nucleus breaks down into two smaller nuclei, releasing energy and more neutrons, typically at least two. These neutrons may hit two more nuclei which, in turn, release more neutrons and the chain reaction goes on fast enough to generate a major explosion out of just a few dozen kilos of uranium. Another good illustration of the power of feedback in complex systems. Then, of course, the chain reaction must stop when it runs out of uranium nuclei, but you may be interested to know that, at the time the first atomic weapon was tested in Alamogordo, in 1945, it is reported that the nuclear physicist Edward Teller feared that the explosion could "ignite the atmosphere" causing the self-sustaining fusion of nitrogen nuclei. That would have destroyed the whole biosphere as a side effect. Fortunately, that did not happen, but note that it was decided to run the test anyway!

Exponential growth is never so strongly desired as in financial systems. Positive interest rates are a basic feature of banks accounts and everyone is happy to see their balance credit grow exponentially. The same is true for debt although, in that case, exponential growth is not supposed to be a good thing. Growth is often the only parameter considered when we read about the state of the national economy. A growing economy, it is said, raises everybody's income and makes society progress toward ever-growing material well being. We are so obsessed with growing the economy that we tend to call "negative growth" what we should more reasonably call "decline," if not "collapse."

The origins of such an emphasis on growth are various, but mostly a modern development. At the beginning of economics as a science, during the 18th and 19th centuries, those larger-than-life figures such as Adam Smith (1723–1790), David Ricardo (1772–1823), John Stuart Mill (1806–1873) and many others, had a strongly pessimistic attitude on the future of the economy. So much that economics was sometimes referred to as "the dismal science." Things changed in time and, with the 1950s, an optimistic mood took over with talks of energy "too cheap to meter", flying cars, and weekend trips to the moon for the whole family. And all that, of course, would have been possible only by letting the economy grow.

It is true that the world's economy has been growing at average rates of the order of 2% over the past century or so, while the human population has been growing, too, although at somewhat smaller rates. But, clearly, growth cannot continue forever. There are several limiting factors, one is that agriculture needs space in order to produce food. Space for agriculture is limited and it even tends to shrink considering the erosion of fertile land and the deplorable human habit of paving the land with concrete. Some people seem to be enthralled by the fact that during the past few decades the world's agricultural food production has been growing faster than the human population, but that does not mean it is a trend that can continue forever. Only some truly exotic technological ideas could overcome this problem, say, finding a way to miniaturize human beings as in the science fiction movie *The Incredible Shrinking Man* (1957). Or, maybe, we could "virtualize" human beings while completely covering the planet with photovoltaic panels [24] so that they would only need virtual food. It goes without saying that these are not especially practical solutions.

Other factors that would limit human growth on a finite planet are the limited mineral resources, in particular of fossil fuels [25]—and not even shale oil is infinite, despite claims to the contrary (as the US ambassador to the EU, Gordon Sondland, said in 2019 [26]) So, what happens when we start running out of the resources that make the economy grow? Or when amoebas run out of food in their Petri dish? Clearly, the answer cannot come from an oversimplified theory that describes growth as a simple exponential function. We need to go more in depth into the mechanisms of growth of complex systems.

Reaching the Limits of the Petri Dish: The Malthus Mode

Sooner or later, the descendants of Amelia the amoeba will have to come to terms with the limited amount of food available in their Petri dish and they will stop growing exponentially. This is a general phenomenon and the first in history to consider it was Thomas Malthus (1766–1834) in his book *An essay on the principle of population* (1798) [27]. According to Malthus, the limited availability of fertile land will eventually stop the growth of the human population because of famines, wars, diseases, or all three together.

Malthus was not a mathematician and he could not do what we call today "curve fitting." All he shows us in his book are tables with data on population and food production in England. He discusses at length how the first (population) grows exponentially while the second (food production) can only grow linearly, at best. A mathematical version of Malthus' ideas came with the work of the French mathematician Pierre François Verhulst (1804–1849) who created an equation that can be seen as representing Malthus' intuition in mathematical terms. We call this equation "logistic," "sigmoid," or, simply "s-shaped." The curve grows rapidly at the beginning—in a nearly exponential manner. Then it tapers down as it approaches the "carrying capacity" of the system, the upper limit of growth. This behavior can be termed the "Malthus mode" of growth (Fig. 2.6).

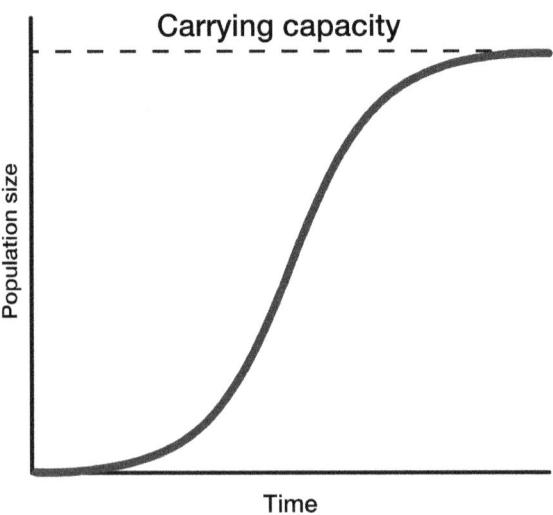

Fig. 2.6 The Verhulst, or "logistic," curve that describes the Malthus mode of growth

You surely know that today Malthus is commonly accused of having made "wrong predictions," so much that the term "Malthusian" or "Neo-Malthusian" is often intended as an insult. Sometimes, Malthus is even accused of having called for the extermination of the poor. For instance, he has been accused of having inspired the British government to exterminate their Irish subjects by artificially creating the great famine of 1845–1849, or at least by doing nothing to mitigate its consequences [28]. All that is far from being the truth: Malthus' "wrong predictions" are just one more example of the many legends infesting the Web. Look at the Verhulst curve: do you see any collapse in it? Of course not. Malthus had no idea of what a "Seneca Collapse" could be like and he never proposed specific dates for future famines. The father of modern catastrophists never was one!

Actually, Malthus was even more optimistic than Verhulst and, in his book, he assumed that it would be possible to keep increasing for many decades the food production of "the island," intended as England. So, according to his prediction, not only there would be no collapse, but the population would continue to grow, just not exponentially. And not just that:

he never even dreamed to say that the poor should be exterminated. The quote often attributed to him that the Irish should be "swept from the land" is a classic case of how the meaning of a text can be twisted by extracting a sentence out of its context. You can read this accusation in Joel Mokyr's 1983 book *Why Ireland Starved* [29]. But if you look at the whole sentence, you see that Malthus was only proposing to industrialize Ireland: "sweeping the Irish from the land" meant that they would move to cities to become factory workers. On the whole, Malthus was, by all means, a man of solid moral principles who did what he could to warn his fellow human beings about a future that he did not like but that he saw as unavoidable if population continued to grow at the rates typical of his times. Given the technologies available when he was writing, the only solution he could propose to avoid that future was sexual abstinence. Not surprisingly, he was not heard.

Malthus ideas were a remarkable step forward in understanding the limits to growth of populations in a closed system, but his model as interpreted by Verhulst is too simplified to be applied to real ecosystems or to human populations. The sigmoid or logistic function turned out to be much more useful in other fields. In commerce, it is typical for the sales of a new product to follow the curve in terms of market penetration. Sales are initially slow, but they grow rapidly as the new product becomes known by potential customers. Then, sales start to fall as the market becomes saturated. Knowing that sales will follow the derivative of the logistic curve gives useful information to producers, telling them about the best timing for marketing a new and improved version with more bells and whistles. There are logistic curves just about everywhere in social and economic systems and the Italian scientist Cesare Marchetti spent most of his career studying them. He reports that [30]

> During the past 30 years, I have analyzed thousands of time series concerning all sorts of social and economic phenomena—from the destruction of the threshers (1 month) to the evolution of British naval power (500 years), from the rounds of artillery shot in Europe by American forces during WWI to American casualties in the Vietnam War, from the victims of the Red Brigades in Italy to those of the witch hunts in the Middle Ages, and so on. The perplexing result is that a very simple logistic model can always fit the data in a predictive format.

Apart from the fact that witch hunts were not a characteristic of Middle Ages but of the later, supposedly "enlightened" Renaissance, Marchetti correctly identifies the general validity of the logistic curve in many human endeavors. It just does not work for ecosystems for a good reason: the fact that living beings tend to die if they have no food. Think of how the logistic curve occurs in chemistry: if you ever performed a *titration* in a laboratory class, you spent your time slowly proceeding to determine the concentration of some element in a solution by gradually building up a logistic curve while adding reactants to the solution, drop by drop. It works beautifully in chemistry because molecules do not "die"—once they have reacted, they just stay there. But that is not the case for amoebas in a Petri dish: they die when they run out of food. So, the limit of the logistic curve is that it never generates collapse. We need something more sophisticated that not only tells us how a system grows, but also how it declines.

What Goes Up, Must Come Down. The Hubbert Mode

Living beings need food and food cannot be infinite in a finite container. When Amelia's daughters start running out of food in their limited Petri dish, they will not follow the gentle, tapering slope that the Malthus/Verhulst model proposes. Rather, they will start dying of starvation. At that point, praying to their Amoeba God won't help them very much to escape their sad destiny. The same is true for an industrial system exploiting a non-renewable resource, say, crude oil. Sooner or later, nothing will be left to extract, at least in conditions that can generate an economic profit. When that point is reached, the industry involved in the extraction must disappear.

The idea that "*what goes up, must go down*" has to be very old. One of its forms is the 1968 song *Spinning Wheel* by the band Blood, Sweat & Tears. But we cannot make a quantitative model out of a song; yet we do need something quantitative and the first model of this kind was proposed by the American geologist Marion King Hubbert in 1956 [31]. It was a model developed not to describe a biological system but an economic system: the extraction of crude oil. The "bell-shaped" curve that Hubbert proposed is sometimes referred to as the "Hubbert Curve." Later on, the idea became known by the term "peak oil" suggested by Colin Campbell in 2001 [32]. We may give this cycle of growth and collapse the name of "Hubbert's mode" (Fig. 2.7).

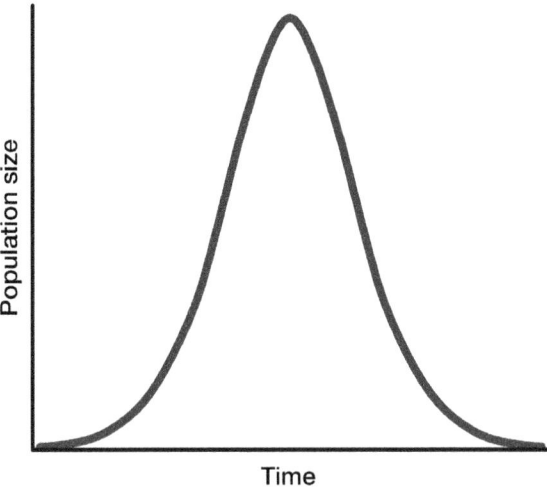

Fig. 2.7 The "Bell Shaped" curve, also called the "Hubbert Curve"

In his 1956 paper, Hubbert presented two projections for the United States oil production. Both showed a peak that was expected to take place around 1965–1970. It turned out to be a good forecast, at least for one of the two curves. The oil production for the contiguous 48 states of the US peaked in 1970, to decline afterward. The curve continued to follow the Hubbert model until the 2000s when investments in extracting oil from shale deposits generated a new cycle of growth. In general, the model can be applied to a variety of systems involving mineral resources, not just oil.

The Hubbert model can also quantitatively describe some economic systems involving renewable resources which cannot be renewed fast enough for their exploitation to continue. A good example is the case of the 19th-century whaling industry. We all probably have a mental model of those times from Herman Melville's novel *Moby Dick*. In the novel, however, Melville never tells us what whale oil was used for: it was used as fuel for oil lamps. In a sense, it was a precursor of modern crude oil. During the 19th century, whale oil turned out to be cheaper than vegetable oils, it was clean and would generate no bad odors when burning. The industry produced also "whalebone," a precursor of modern plastic. It was used for such applications as corset stiffeners for ladies and back scratchers. As a consequence, whaling went through a phenomenal growth cycle that transformed it into a major global industry.

Around 1850, the American whaling industry was at its peak, producing a total that arrived close to 15 million gallons of oil per year (Starbuck 1878).

But decline soon started and the production curve of both whale oil and bone followed a bell-shaped Hubbert curve (Bardi 2004) (Fig. 2.8).

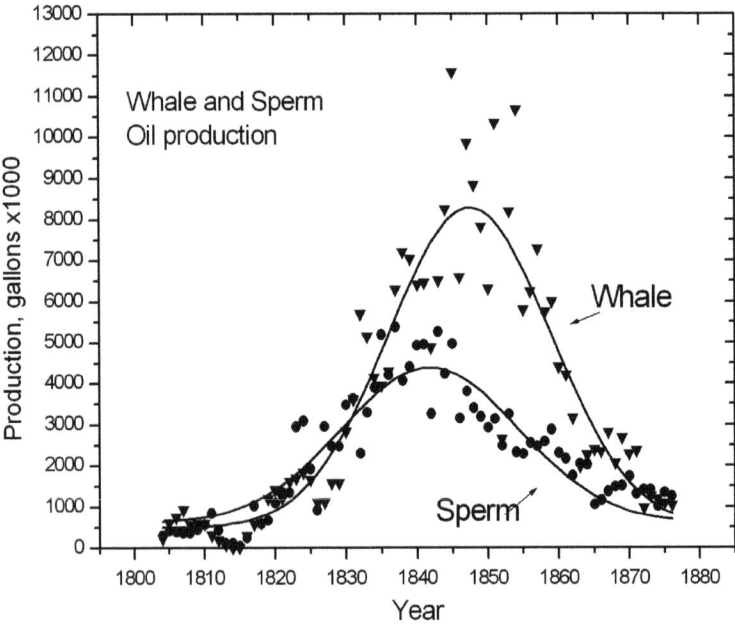

Fig. 2.8   The cycle of the American whaling industry in the 19th century. Here, "Whale" stands for the "right whale" species and "Sperm" for the more prized "Sperm Whale". Data from Starbuck's History of the American Whale Fishery (1878)

A common interpretation for the decline of whaling is that it was caused by the introduction of kerosene as lamp fuel, something that destroyed the whale oil market. But that means stretching the dates more than a little. When the first oil well was drilled in Pennsylvania, in 1859, the American production of whale oil had been in decline for years and it was half of what it had been at the peak. It took several more years for kerosene to overtake whale oil in terms of production. Actually, you could reasonably reverse the cause and effect chain: kerosene was marketed *because* the production of whale oil was declining!

Besides, kerosene was considered an inferior fuel, and there always remained a market for whale oil as lamp fuel. It was more expensive but considered to be of better quality. To say nothing about the fact that the Pennsylvania wells could not produce good back-scratchers—at least until it was learned how to make plastics out of hydrocarbons but that took almost a century.

Overall, the reason for the decline of the whaling industry is clear: whales were killed at a much faster rate than they could reproduce. According to some studies, at the end of the 19th century there remained in the world's oceans only about 50 females of the species most commonly hunted by whalers, the "right whale" [33]. Whales turned out to be a nearly non-renewable resource. Still today, the number of right whales remains small and the species is at risk of extinction even though it is not hunted anymore—at least officially [34].

At this point, we must explain why the Hubbert model works so well in so many cases for both renewable and non-renewable resources (although not all cases, of course). A problem, here, is that Hubbert was not a theorist in the sense of being a person who built mathematical models. He was an empiricist who looked at the data and used common sense to interpret them. From his 1956 paper [31], it is clear that he looked first at the historical oil production data for Ohio and Illinois, noting that in both cases the production curve was approximately bell-shaped. Then, he assumed that the production for the whole conterminous US states would follow a similar curve. It would have to be subjected to the constraints of a limited total amount of extractable oil and on the fact that the extraction rate must be zero at the beginning and at the end of the production cycle. Given the approximation involved in this approach, he did not need to go into the details of what kind of mathematical function would provide the "best" bell-shaped curve. Indeed, the curves of the figures in his early papers look as if they had been drawn by hand.

Later on, the Hubbert curve was identified with one of the several bell-shaped curves existing in the arsenal of statistics. The preferred one was often the derivative of the logistic function as you can find described in a paper by French oil expert Jean Laherrère [35]. Other bell-shaped curves exist under names such as the Gompertz, the Ogee, the Bass, and others. Of these, only the Bass function has been used as a tool for interpreting resource depletion [36]. And, of course, the paradigmatic bell-shaped curve is the "Gaussian" or "normal" curve, which can also fit historical production curves [37].

But why should production follow a bell-shaped curve? For instance, in the case of the Gaussian, the bell shape is generated by a large number of independent events occurring at random, as happens when you roll dice. But you would hardly be able to define the extractive cycle of the oil industry in terms of rolling dice. A persistent interpretation of the Hubbert curve derives from a statement by Campbell and Laherrère in 1998 [38] who noted that "adding the output of fields of various sizes and ages usually yields a bell-shaped production curve for the region as a whole." But this simply an observation, not a model.

So, we need a better theory and we can create one using the tool called "system dynamics." Without going into the details of how it works, let us just say that system dynamics is based on the idea that energy (or some other quantity) flows from one "stock" to another. Then, the formalism of the theory provides tools to describe a system in terms of the various stocks and flows it is composed of. In the simplest case, the stock could be a bathtub partly filled with water—it may gain water from the tap and lose it from the sink. Some experts of system dynamics seem to be fond of the bathtub analogy and tend to use it extensively as a teaching tool [39] (Fig. 2.9).

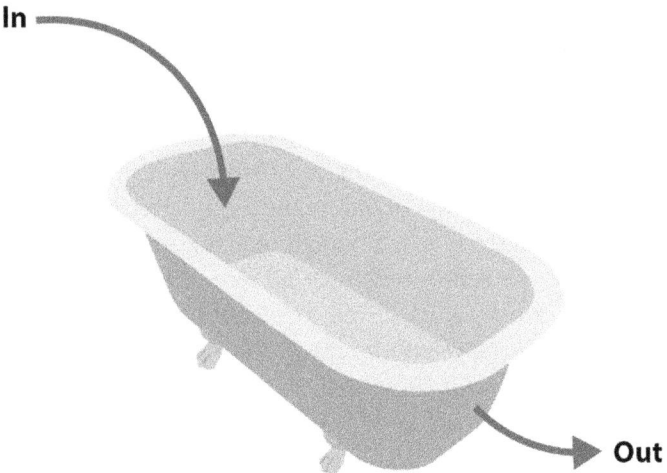

Fig. 2.9 In system dynamics, a bathtub is often taken as an example of the general concept of "stock."

An oil well or a whole oil field can be considered as a stock of oil. There are a few differences with respect to a bathtub: one is that there is no "faucet" to replenish the oil stock: oil is a non-renewable resource. Another difference is that the sink is not just a plug that can be removed at will. In order to be extracted or "produced," as is commonly said in the jargon of the oil industry, one needs equipment, people, machinery, transportation, financing, and more. This need can be represented in terms of a second stock of the model that we can dub as "capital:" it is the ensemble of the material, human, and financial resources needed to extract oil. This second stock can be seen as created by the first, although indirectly. With the profits from extraction, the oil industry can invest in more capital that can be used to extract more oil. The two stocks are in a classic relation of enhancing feedbacks with each other. The more capital you have, the faster you extract the oil, the more oil there is, the faster it can be extracted. Then, of course, there is an unavoidable damping feedback that's created by oil depletion. As you gradually empty the oil stock, profits go down, too, and that negatively affects the extractive capital that tends to decline because of depreciation. Eventually, these factors stop the growth of production and cause it to start declining. In the long run, no matter how much capital you can manage to generate, you cannot extract oil that is not there, and production must stop. The model generates a very nice "bell-shaped" curve that corresponds to the Hubbert one.

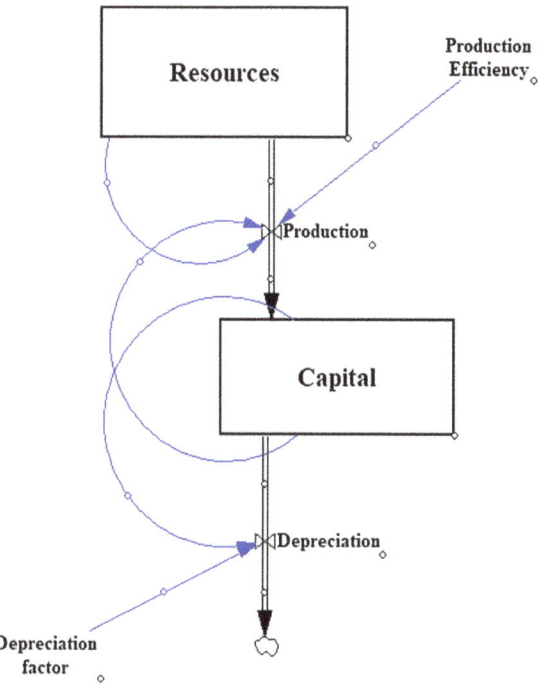

Fig. 2.10 A simple representation of the hubbert model according to the conventions of system dynamics. Rectangles represent stocks, double sided arrows represent flows. Thin arrows represent feedbacks. Graphic created using the Vensim (TM) software

In the figure, you can see a simple system dynamics representation of the Hubbert model (Fig. 2.10). Here, we just need to note how stocks are represented by boxes and flows by means of thick or double-line arrows. The feedbacks of the system are represented by thin or single-line arrows acting on the "valves" of the system. The drawing does not tell you how exactly the action is performed, that needs to be specified by the modeler "under the hood." The program provides tools to input equations that tell the program that, for instance, the flow out of a stock is proportional to the size of that stock. But it might be proportional to other variables, or be subjected to thresholds, or the like. In any case, the concept is clear: the system evolves and it dissipates energy as a function of the amount of resource available. For details about how these models can be used to describe collapse, you can see my first book on the Seneca effect [10] (this model, incidentally, is a simplified version of the "Lotka-Volterra" model, well known in biology).

Note also that the model contains *two* main parameters (or stocks), one for the amount of the resource and one for the amount of capital available. The second parameter is not visible in the standard version of the Hubbert model. Yet, surely oil does not extract itself, so the system dynamics version of the model tells us something important about how depletion works: you do not just "run out" of a natural resource—you may (and you do) run out of the capital resources needed to exploit the resource as the result of increased costs of production and lower profits. Eventually, that discourages new investments and your capital resources waste away in the phenomenon called "depreciation." It is often reported that Ahmed Zaki Yamani, former minister of petroleum in Saudi Arabia said that "The stone age did not end because people ran out of stones, so the oil age will not end because people will run out of oil." Indeed, the oil age will end because people will run out of the capital needed to extract oil: who would ever invest in something that does not provide a profit?

This model of the Hubbert curve is still highly simplified, but it is sufficiently realistic to be able to provide a quantitative description of several real-world systems, fisheries for instance [40]. But many systems show more rapid collapse than growth, in other words, the Hubbert model does not normally generate the "Seneca Curve." So, we need more sophisticated models.

The Way to Ruin is Rapid: The Seneca Mode

When the daughters of Amelia the amoeba start running out of food in their Petri dish, their destiny is unavoidable: they must die of starvation. Even turning to cannibalism (amoebas are not known to be fussy about what they eat) will not change their fate. They will also have problems with their own excreta, that is, they will start being poisoned by it. Whatever may happen to the poor critters, it is clear that dying is much faster than the laborious process of being born, growing, and reproducing. So, we do not expect the population curve of the amoebas to be the symmetric "bell-shaped" Hubbert curve. It will be an asymmetric "sawtooth-shaped" curve with population declining faster than it grew.

The Roman philosopher Lucius Annaeus Seneca was possibly the first to understand that complex systems tend to grow slowly and collapse rapidly, therefore, we can use the term "Seneca Mode" for this kind of mode. Seneca was not a mathematician, at most he had some knowledge of geometry but no idea of what an "equation" is, and he did not even know our modern Arabic numbers. And it goes without saying that he could not draw a Cartesian plot. Yet, he had a good qualitative grasp of what we call today a "complex system." After all, our mind is also a complex system and intuition is not a bad way to understand the world of complex systems. It is the theme that I call "Seneca Effect."

A real-world example of a Seneca collapse is that of the population of the reindeer of St. Matthew Island in the Northern part of the Pacific Ocean. In 1944 the US military brought there a small group of reindeer, 29 of them, with the idea of using them as "meat storage on legs," without the need of refrigerators. The idea seemed to work, the reindeer ate lichens and grass and they reproduced in great numbers. The enhancing feedback of growing populations had kicked in and, some 20 years later, in 1963, the reindeer population had swollen to about 6,000 individuals. Too many for that small island: they were running out of food and not just that: they were sick. This is the effect I call "dynamic crunch," it occurs when more than a single factor start to operate in a complex system to bring the whole thing down, fast. With almost no fat left on their bodies, the reindeer were not only starving but also vulnerable to infections and ticks. Then, they started dying. A couple of harsh winters finished the job: by 1968 there remained just a small group of emaciated and parasite-infested females. Two years later, there were no reindeer alive left: only their bones strewn all over the island [38] (Fig. 2.11).

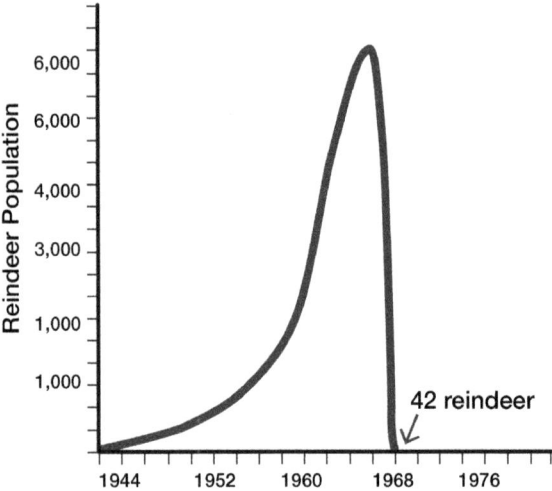

Fig. 2.11 The reindeer population of St. Matthew Island. Data from [38]

Let us see another example of a biological population wiped out in a Seneca-like collapse. It is the crash of the sturgeon fishery in the Caspian Sea. The collapse of fisheries is a common effect of overfishing [41] as we saw in an earlier section about the Atlantic cod fishery. But the case of the sturgeon is special in the fact that it shows such an abrupt decline. As you know, the sturgeon is the origin of the prized black caviar, today practically disappeared from the world market. It was replaced, but only in part, by caviar obtained from "farmed" sturgeon (data from FAOSTAT) (Fig. 2.12).

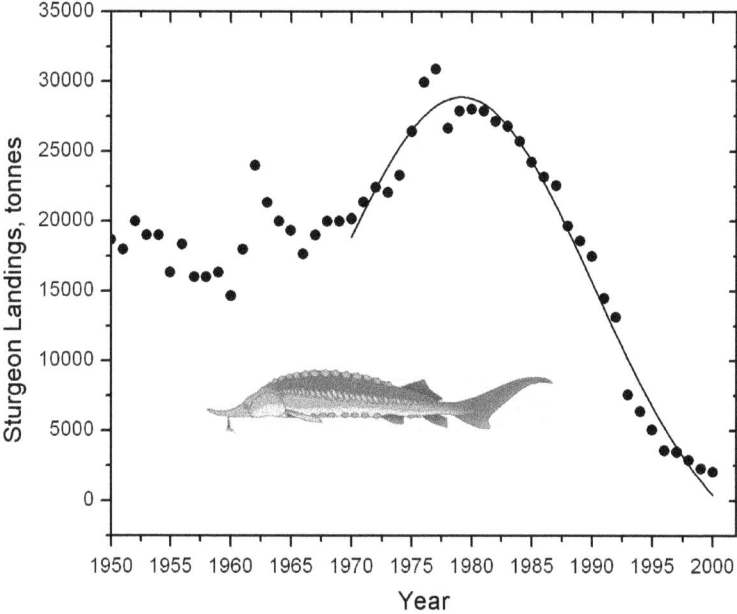

Fig. 2.12 The collapse of sturgeon fishery in the Caspian sea. Data from FAOSTAT

The data available on the sturgeon fishing cycle are only partial but, clearly, multiple factors played a role in the process. In particular, the Caspian sturgeon population was affected by the pollution generated by the industries that used the sea as a dump for their waste. Another factor was the high value of caviar, for which people were willing to pay a price wholly disproportionate to its nutritional value. This factor may well have led the industry to an illegal overkill of the sturgeons, which could not be managed by regulatory agencies. The fall of the Soviet Union, in progress when the collapse started, made it impossible to manage Caspian fishing in a centralized way and to suppress illegal activities. We have, again, a dynamic crunch effect: two or more factors collaborating to bring the system down.

The Seneca effect takes place in all kinds of systems, not just biological trophic chains. We can have industrial or financial chains forming a complex system where the bottom element of the chain is the market for a certain product. The saturation of the market takes the role of the depletion of natural resources and a parallel factor is financial: at some moment the banks may refuse to finance a company that shows signs of decline. That quickly finishes the system off. We shall see examples of Seneca-like financial collapses in later chapters. Here, let me report the case of the Western Roman Empire as a large-scale example. You see in this image some data reported by Taagepera [42]. These are partial data because they refer only to the extension of the empire but are representative of the phenomenon of growth and collapse of a large social system (the Y scale is in million square km) (Fig. 2.13).

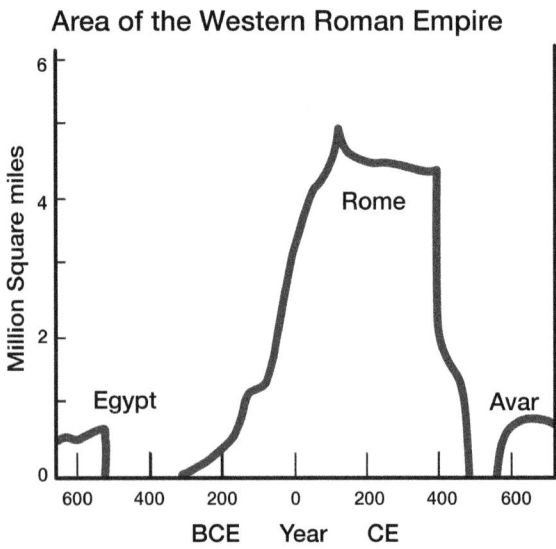

Fig. 2.13 The Collapse of the Roman Empire population, data from Taagepera's "An Analysis of the Surface Area of the Western Roman Empire until CE 476", 1968

There exist more data showing that the Roman collapse was rapid and brutal. For instance, archaeological data show an abrupt decline of the pollution arising from the Roman metallurgical industry. What we see is a true collapse of the Roman industry and, with it, of the whole Roman society [43].

Now, do these collapses have something in common? Yes, they do. They are systems where more than one factor contributed to the decline: it is the "dynamic crunch" effect. There is a characteristic pattern: negative factors tend to gang up to hasten the fall, as if there were some evil mind busy at planning the disaster. But it is not the case, it is simply that in complex systems all the elements tend to be connected and, when one fails, it takes other elements with it: it is the typical avalanche mechanism. In a complex system there are always more than a simple chain of cases of cause and effect. Collapse is a complex phenomenon, typical of complex systems, and it does occur for a complex combination of effects.

Could we describe the Seneca collapse phenomenon by means of a mathematical model? There are several functions that can provide a forward-skewed cyclical curve, but it is difficult to find one that embodies the factors that I just described. If the collapse is linked to multiple factors, the best model is one that explicitly describes these factors. So, a good approach is to use system dynamics, as done in the previous section. In the Hubbert model, the model takes into account two parameters: resources and capital—the latter is the entity (industry or population) that grows on exploiting the resource. There is only one constraint, here: the gradual depletion of the resource stock that makes production more expensive. Is it possible to modify this model adding more constraints?

Yes, it is possible. In a model that I developed in 2011, I added one more equation and one more parameter to the model [44], defining it as "pollution." It does not have to be identified with a specific kind of pollution, it is just one more entity that negatively affects the "population" or the "capital" parameter. It might be invading barbarians, terrorism, climate change, social unrest, epidemics, and more. The extra constraint slows down the growth of the system and, eventually, accelerates its collapse. This is a very general behavior of complex systems. When they are multi-parameter, it normally happens that several parameters "gang up" together to bring the system down. And this, typically, generates the Seneca Cliff.

As far as I know, the three-parameter model is the only one generating the Seneca cliff that is completely self-contained, that is, it contains only "endogenous" parameters. But it is not the only possible model for the Seneca effect. Assuming external factors ("forcings"), it is possible to generate the cliff with just two parameters (a "two-stock" model). Donella Meadows shows such a model in her book *Thinking in Systems* [45]. In this case, the model includes an external constraint on the system that keeps growth below a fixed maximum rate. As a result, at some point the system ceases to grow and goes into free fall, generating the cliff. Another model that generates collapse has been proposed by Francois Roddier and describes the Seneca cliff as related to the Van Der Waals surface, a physical phenomenon related to the condensation of gases [46]. There also exist qualitative models accounting for systemic collapse: one is "Gause's law of competitive exclusion," well known in biology. It says that when two species compete for the same resources, one will rapidly go extinct. There are several examples in biology, but also in economics. One is how the diffusion of Web-based advertising had the effect of starving newspapers of their advertising revenues, causing their collapse in terms of sales and profits [47].

It is also commonly observed that trying to avoid or to slow down collapse may lead to a more rapid collapse after some time. Per Bak reports some examples in this sense in his book "*How Nature Works*" (1996) [14]. In the section titled "Replaying the Tape of Evolution," Bak describes how they intervened on the evolution of a system that simulated the evolution of a species in an ecosystem, trying to eliminate one of the largest fluctuations (p. 157):

> We then identified the event that initiated one of the larger avalanches involving that particular site. Of course, that could be done only in hindsight… We eliminated that event by replacing the fitness with a higher value and thus preventing extinction there. This interruption could correspond to changing the path of a meteor, or preventing the frog from developing its slippery tongue. We then ran the simulation again. .. At the point where the minor perturbation was made, history changed .. the larger punctuation is gone. However that did not prevent disasters at all. Other perturbations happened at later points. Thus large perturbations cannot be prevented by local manipulation in an attempt to remove the source of the catastrophe. If the dinosaurs had not been eradicated by a meteor (if they indeed were), some other large group of species would be eliminated by some other triggering event.

As another example, in his book *Structures* (1968) [48], James Gordon reports how the attempt to reinforce a mechanical device may result in weakening it. One of the examples he makes is the Fokker D8 *Eindecker* fighter plane of the First World War. It tended to lose its wings under stress and the initial attempts to reinforce them in the wrong places led to even worse problems of structural weakness. The builders could solve the problem by *weakening* one of the beams of the wing in order to distribute the load more equally. Surely a counterintuitive idea, but it worked.

No doubt, there are many other ways to develop qualitative and quantitative models where multiple effects, external or internal, combine to take down the system and generate a rapid collapse. One is the model used for the well known 1972 study, *The Limits to Growth* [49] (Fig. 2.14).

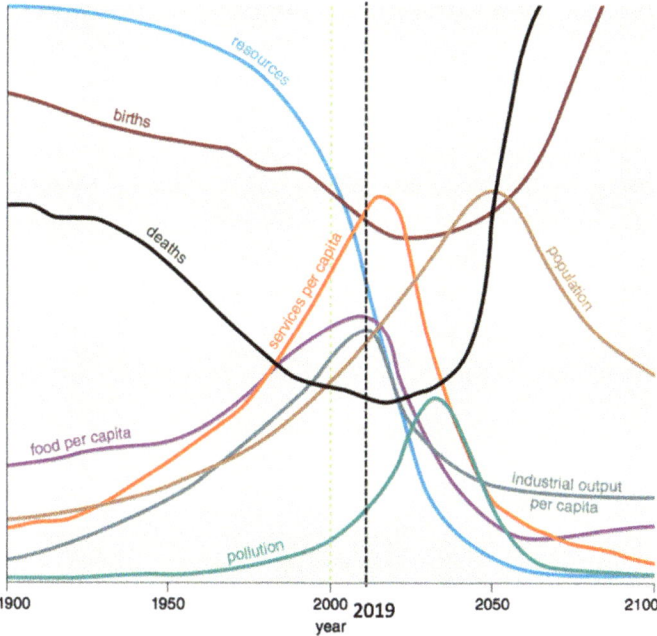

Fig. 2.14 The "Base Case" scenario of the 1972 edition of The Limits to Growth. Data courtesy of Mr. Dennis Meadows

If you look carefully at the curves of the base case scenario (shown in the figure) as well as of those of other scenarios presented in the study, you will see that, in many cases, there are cycles where the decline is faster than growth. They had already discovered the Seneca Effect even though they did not use that name.

Overall, none of these models prove that the Seneca Effect must necessarily happen in all complex systems, but it is clear that it is a common phenomenon. It does not seem to matter whether we are amoebas or human beings, we live in a world were collapses are not a bug, they are a feature.

When the Unexpected Strikes: The Hokusai Mode

The daughters of Amelia the amoeba are supposed to live and die in their Petri dish, facing all the opportunities and challenges that they find in there. But it cannot be excluded that something completely different could happen to them. Imagine, for instance, that a bored graduate student of the lab decides that the experiment has not worked the way it should have, and proceeds to wash and disinfect the Petri dish. That would be a sudden and unpredictable catastrophe for the amoebas, mercilessly wiped out in a very short time.

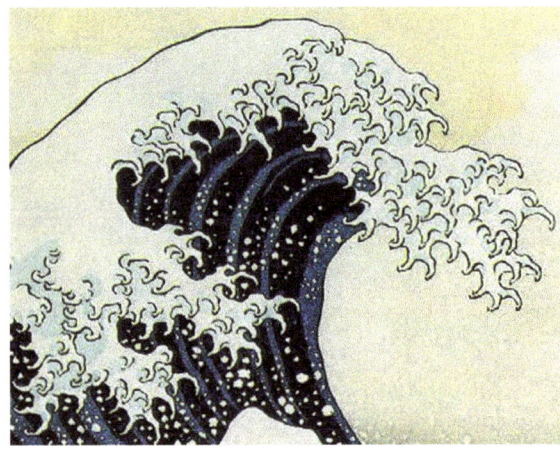

Fig. 2.15 A detail of the Great Wave of Kanagawa a print by Katsushita Hokusai published sometime between 1829 and 1833 in Japan

The Japanese painter Hokusai never thought in terms of mathematical models but he provided a great visual impression of what happens when things get truly bad all of a sudden with his famous painting, *The Wave* (ca. 1830). Most likely, the painting represents a *tsunami*, a Japanese word meaning "harbor wave." It is the result of a sudden rise of the sea level that reaches the coast, normally following an earthquake. When a tsunami strikes, not only is decline faster than growth, but it is so fast and gives you so little warning that there is little you can do to avoid it. For instance, when the Tōhoku Earthquake struck Japan, on March 11th, 2011, the best that the alert system could do was to announce the arrival of the great wave little more than half an hour in advance. Many people were taken by surprise and swept away, the number of casualties ran into more than 15 thousand. We could call this kind of sudden and unexpected collapses the "Hokusai Mode. (Fig. 2.15)"

The characteristic of Hokusai-mode collapses is not so much their magnitude, although they may be large. It is that, in statistical terms, they are outside the range of the events you normally take precautions against. Suppose you are walking in town: you know that it involves a certain danger of being hit by a car and you minimize it by walking on the sidewalk and carefully watching left and right before crossing the street. Yes, but your precautions are useless if you are hit by a falling tile from a building. It is a rare event, so rare that there do not seem to exist statistical data on its frequency worldwide. But it does happen: for instance, in 2017 a tourist from Spain was killed by a falling stone while visiting one of Florence's ancient churches, the Basilica of Santa Croce [50]. It was truly a sudden collapse for this man and it is impressive to think that he had come all the way from Spain to find himself exactly at the place and at the moment when that stone fell. Was there some logic in the universe that established that the trajectory of this particular stone and this particular person had to intersect? We cannot say, except by noting that if a stone falls inside a church normally full of tourists, it is likely that it would hit one of them.

Probabilities can change depending on how we frame their calculation. Call it an "act of God" or simply bad luck, being hit by a falling stone is one of those events that can strike people without warning and without any possibility to be prepared in advance for them. The only way to be safe from stones falling from the roofs of ancient buildings is to never visit one, but that is hardly practical and surely not justified. Every year, millions of tourists visit ancient churches in Florence and the 2017 incident seems to be the only one of this kind reported in recent times.

A Hokusai-mode collapse needs not to be completely sudden: it may give you some advance warning, but not enough to do something to avoid the disaster damage. An especially impressive example is related to the cannon balls fired during the battles of Napoleonic times. Donald Graves notes that,

One characteristic of roundshot was that, when it bounced along the ground, it often appeared to be moving slowly. Unwary soldiers sometimes tried to stop it with their feet only to suffer an amputation as clean as that performed by any surgeon [51].

If you were a conscript of Napoleon's or Wellington's army on 18 June 1815, there was little you could to do except trusting your luck to avoid being hit by a roundshot and becoming part of the statistics of the casualties of that day. Another example of a situation where the best strategy is not to be there.

In other cases, the advance warning time may be much larger than that given by a bouncing cannonball, but still insufficient to do anything to avoid the disaster. Consider the risk of the fall of large meteorites. Almost unknown to the public, there exists an early warning system for incoming meteorites managed by NASA. It is called ATLAS (Asteroid Terrestrial-impact Last Alert System) [52] and is reported to be able to provide "*a one day's warning for a 30-kiloton "town killer," a week for a 5-megaton "city killer," and three weeks for a 100-megaton "county killer."* With this system, we are in a classic science fiction situation: suppose the system returns the information that New York is going to be hit by a 5 megaton meteorite in a week, what could the authorities do? Evacuate millions of people in a week? And where would they go? More than that, would the authorities believe the scientists launching the warning or would not they rather accuse them of "alarmism," of peddling catastrophes in order to gain prestige and research grants? Yet, cities have been obliterated in history, if not by meteorites, at least by equivalent amounts of energy generated by nuclear weapons. In 1945, the US forces destroyed the Japanese cities of Hiroshima and Nagasaki by the first nuclear bombing in history. Could that happen again? Probably yes, but two cases are not enough to create a statistical distribution.

Hokusai-mode disasters are both rare and unpredictable, but that does not mean we cannot say at least something about their statistical properties. Some might be termed "black swans," according to the description given by Nassim Taleb [53]. These events are part of known statistical distributions, but extreme in terms of their large size and low probability. For instance, the 2011 Tōhoku Earthquake was very large, among the largest ever observed, but it was not outside the known size distribution of earthquakes. Its probability could be calculated, at least approximately, but it was considered to be so low that no precaution was taken for the height of the associated tsunami waves. This is, indeed a black swan: a low probability event that is normally discounted but that can happen, and does happen.

Other Hokusai-like collapses are not just extreme, but fully outside the boundaries of the system in terms of the typical power laws that govern this kind of phenomena. Sornette coined the term "Dragon King" when he noted how the size of the city of Paris is so large that it does not fit the size distribution law that connects all other French cities [54]. Previously, Jean Laherrere had noted the same phenomenon, but he had been using just the term "King" [55]. Of course, Paris is not a collapse, but it is a good example of an anomaly in a power law probability distribution. Catastrophic examples of Dragon King events include the largest radiation release events occurring in nuclear power plant accidents, such as at Chernobyl, the largest crashes in financial markets, and some wild oscillations of market prices [54].

There do not seem to exist mathematical models for Dragon Kings of the kind that describe, for instance, the Seneca curve. There are equations that can generate a curve that looks a little like the Hokusai wave, or the shape of any wave crashing on the beach. One is the Duffing oscillator [56], a non-linear, periodically forced oscillator. It is an interesting mathematical exercise but not something that can be used to predict catastrophes in the real world. The only thing we can say is that the universe does not move smoothly. We know that it tends to dissipate entropy as fast as possible and that it does that in bursts. Some of these bursts are very large and are aptly termed "Dragon Kings." For the Chinese, a dragon is a benevolent entity, often bringing rain. For the Westerners, it is an evil fire-breathing creature. Perhaps the principle of maximum entropy dissipation is the result of the principle of Yin and Yang, that all things exist as inseparable and contradictory opposites. And thats, again, is the way the universe works.

Life After the Cliff: The Seneca Rebound Mode

So far, we have been thinking of Amoebas in a Petri dish as a strictly limited system in terms of available resources. The daughters of Amelia may grow, peak, and then decline following various trajectories, but their destiny is unavoidable: when they run out of food, they die—with the additional chance of suffocating in their own excreta.

But let us assume that the resources are not really a one-go; instead, let us imagine that the experimenter is benevolent and merciful and does not want the amoebas to die. So, she keeps nourishing the little critters in their Petri dish and also provides a way to flush away their excreta. Or, maybe, amoebas could find themselves living in one of those autonomous ecosystems that can be created inside a sealed glass container [57]. These containers are mainly decorative objects, but the creatures inside are reported to be able to survive for some years. In this case, the destiny of Amelia's daughters need not be apocalyptic anymore: they may survive if they can adapt to the limited amount of available food and space. That does not mean they won't have a hard time in terms of population peaks and valleys as they periodically overshoot the availability of food, their population crashes, only to restart when food is accumulated again. In biology, this kind of cycles is called the "Lotka-Volterra" (LV) mechanism, from the names of the scientists who independently proposed a mathematical model for the growth of simple biological and economic systems [58, 59]. Since the model sees the return of growth after a crash that may take the Seneca shape, we may also call it the "Seneca Rebound."

The Lotka-Volterra model is often referred to as the "prey-predator" model, but it is also known as the "foxes and rabbits" model. It is curious to note that neither Lotka nor Volterra ever thought in terms of rabbits and foxes, nor of other specific prey-predator couples. Lotka's model was expressed in a very abstract and general way, related to the thermodynamics of living systems. Volterra, instead, developed his model as a way to explain the behavior of the fish populations of the Adriatic sea as it was reported by his son in law, a marine biologist who studied the Italian Adriatic fishery during the First World War. The destiny of the ideas of Lotka and Volterra was similar to that of Jean Baptiste Lamarck's ideas on evolution, often reported as having to do with the neck of giraffes, but Lamarck never mentioned giraffes!

Call it the way you like, the Lotka-Volterra model is very similar to the Hubbert model that we saw in an earlier section. The only difference is that in the Hubbert model the resource (oil) is supposed to be non-renewable, whereas in the Lotka-Volterra one the resources (rabbits) can regrow after it has been consumed. The result is that the "bell-shaped" production curve does not end at zero, but picks up again and again in an infinite series of oscillations. The predator population, (the foxes), will periodically consume their resources (the rabbits) so fast that the predators will starve and die. Then, with so few foxes around, the rabbit population will restart growing. And then, the numerous rabbits again become an easy prey for the growing fox population, and so it goes, forever, at least in the model (Fig. 2.16).

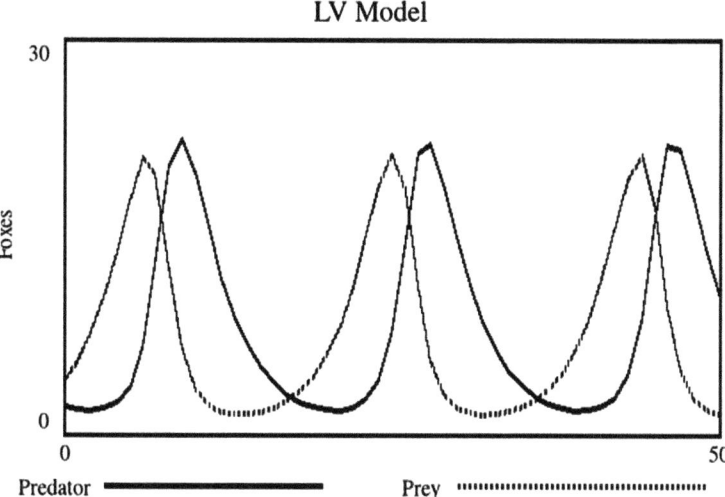

Fig. 2.16 Typical oscillations in the populations of a coupled prey-predator system according to the Lotka-Volterra model

In the figure, you see a typical example of the results of the model. In the simplest version of the model, these oscillations go on forever, unchanged. Note the "bell-shaped" curves for both populations but, also, how the curves for the prey (the rabbits) decline faster than they grow. This is one of the manifestations of the Seneca Effect, it is a typical behavior of a population subjected to predation. Note also that the foxes show a different behavior, declining slowly. That's because the trophic chain in the model is formed of two steps only—that is, the predator has no predators. So, foxes die slowly by starvation rather than fast by being eaten by a growing population of predators, as happens instead to rabbits.

The problem with the Lotka-Volterra model in this simple form is that no wild population is known to behave in this way [60]. The model was found to be able to describe the behavior of real populations only in very specific and controlled laboratory conditions, with micro-organisms such as yeasts and bacteria. In some textbooks, you may be told of the case of the lynx and the snowshoe hare, populations which can be roughly measured by means of the data on the number of pelts brought by hunters to the Hudson Bay Company, in Canada in 1845–1935. A more recent study dealt with the moose and wolf population in a park in British Columbia, Canada [61], These cases do show oscillations, but the agreement with the model is far from being satisfactory.

All that does not mean that the Lotka-Volterra model is useless. We know that the map is not the territory and the model is not the real system. The LV model was never supposed to describe real ecosystems, but it was found to work very well for some economic systems, such as for fisheries [41]. That is because the human industrial ecosystem is often much simpler than a biological ecosystem. An entity such as the fishing industry does not have the incredible tangle of complex interactions existing in natural ecosystems (Fig. 2.17).

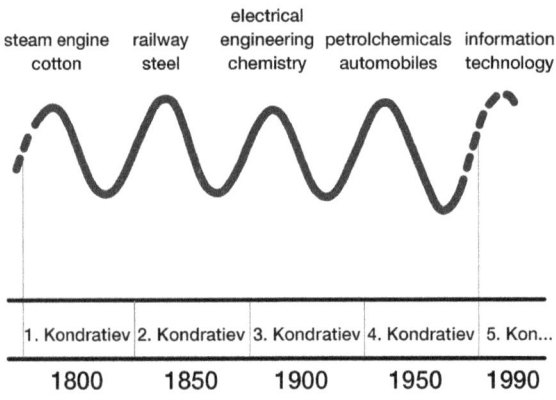

Kondratiev Cycles

Fig. 2.17 Schematic behavior of Kondratiev waves in the economy. Adapted from Rursus, https://en.wikipedia.org/wiki/Kondratiev_wave#/media/File:Kondratieff_Wave.svg

History is full of economic cycles and collapses: economists have played with the matter a lot and we can list at least seven modes of cycling that have names: (1) Kondratiev, (2) Kuznets, (3) Schumpeter, (4) Kalecki, (5) Goodwin, (6) Kaldor, and (7) Minsky, (8) Kitchin, (9) Juglar [62]. Of these, probably the best-known variety is the Kondratiev wave, from the name of the Russian economist Nikolai Dmitriyevich Kondratiev (1892–1938). He noted the long waves with a period of somewhat more than 54 years that today take his name and which are commonly attributed to technological progress. That is, each new invention generates a new cycle of growth that then subsides when the market is saturated, to restart as the result of a new invention.

Kondratiev's ideas were popularized by Schumpeter in *The Theory of Economic Development* [63], and are well known today. There is no doubt that these cycles exist, but their interpretation as the result of technological improvements is at least debatable—why should progress follow a cycle? And why a regular 54 year cycle? Similar considerations hold for all the other cycles that form a remarkable fauna of studies and interpretations. As an example, here are some data for the cycle of cement production in Italy (data source, AITEC—www.aitec.it) (Fig. 2.18).

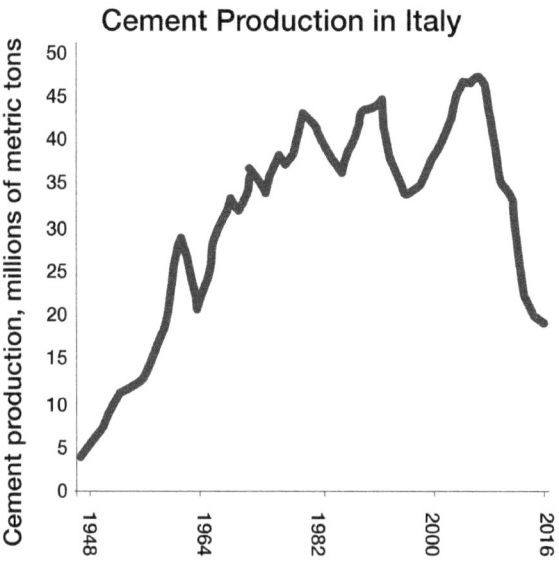

Fig. 2.18   Cement production in Italy in thousands of tons. Data from www.aitec.it

The curve can be interpreted in terms of three main production cycles, about 12 years long, with the first peak in 1982, the second in 1994, and the third in 2007. Ten years later, in 2017, a further cycle of growth may be starting. So, it may be that the building construction market in Italy is subjected to long term cycles of variable length spanning the past 40 years or so. But note that there has been no evident technological improvement in the ways cement is made, nor in how buildings are constructed, that could explain these cycles. Here, we are seeing only economical and market factors at play.

These cycles could be qualitatively interpreted using models similar to the Lotka-Volterra one where we could interpret the industry as the "predator" and the market as the "prey." What happens is that the industry exploits the market to grow, but in doing so it tends to saturate the demand. With lower demand, profits fall, investments decline, and—eventually—the whole industry shrinks. But, with time, demand returns in a market that has been in decline for a while, and the cycle restarts. It is, in the end, no different from the case of Foxes and Rabbits. It may well be that the same economic factors, rather than technological ones, are at the basis of the Kondratiev cycles.

Economists rarely mention the Lotka-Volterra model in their papers. That's because science, as it is today, is strongly compartmentalized and no scientist wants to expand outside his specific area of competence. In this case, then, there is an ongoing conflict between standard economics and the growing area called "biophysical economics" or "econophysics." Traditional economists resent what they see as an invasion of their field on the part of physicists, and tend to make choices that show to everyone which banner they are following. Nevertheless, differential equations of the kind at the basis of the Lotka-Volterra are used in economics (see e.g. Clark and Munro Model for the overexploitation of fisheries [63]).

Overall, we can say that cycles of growth and decline exist in both ecosystems and economic systems. It is, again, part of the way the universe works: what goes up must go down, and what goes down may go up. In system science, we may say that a complex system may "orbit" around an attractor in a way similar to a planet orbiting around a star. So, we may expect these cycles to affect our lives and to take us from phases of optimism to phases of dismay and back.

The problem with these cycles is that it would be a big mistake to use them to make predictions of the future. They are the result of the interplay of internal factors of the complex system which is the economy, but they may be affected by external forcings that may disrupt the tendency of the system to keep staying around the attractor. So, apparently stable economic cycles may be disrupted by non-market factors acting as forcings: climate change and resource depletion, for instance, may destabilize the whole economic system and send the various economic bodies that compose it crashing against each other, in what might take the shape of wars.

So, take into account that economic cycles exist, but do not plan your activities on the basis of the idea that they will keep going with the same frequency and intensity. As an example, the Italian building industry is right now (2019) hoping that a new cycle of growth will bring the cement market back to what it was at its apex, in 2007. It might, but—as usual—whether you are an amoeba or a man, the best way to make wrong predictions is to assume that the future will be like the past.

References

1. De Shong Meador, B.: Princess, Priestess, Poet. University of Texas Press (2009)
2. Black, J.A., Cunningham, G., Fluckiger-Hawker, E., Robson, E., Zólyomi, G., Inana and Ebih translation.: The Electronic text corpus of sumerian literature. http://etcsl.orinst.ox.ac.uk/section1/tr132.htm. Accessed 3rd Aug 2015
3. Bardi, U., Inanna, H.A., Ebih, R.L.: A report of an ancient ecological catastrophe? (2015). http://chimeramyth.blogspot.com/2015/08/inanna-and-ebih-report-of-ancient.html. Accessed 17th Apr 2019
4. Prigogine, I.: Symmetry breaking instabilities in dissipative systems. II. J. Chem. Phys. **48**, 1695 (1968)
5. Hall, C.A., Cleveland, C.J., Kaufmann, R.: Energy and Resource Quality: the Ecology of the Economic Process. Wiley Interscience (1986)
6. Odum, H.T.T.: Self-organization, transformity, and information. Science **242** (80), 1132–1139 (1988)
7. Raugei, M.: Net energy analysis must not compare apples and oranges. Nat. Energy **4**, 86–88 (2019)
8. Sgouridis, S., Bardi, U., Csala, D.: The sower's way. quantifying the narrowing net-energy pathways to a global energy transition (2016). arXiv preprint arXiv:1410.7172
9. Roddier, F.: Thermodynamiqe de l'évolution. Parole (2012)
10. Bardi, U.: The Seneca effect. Why Growth Is Slow but Collapse Is Rapid. Springer Verlag (2017)
11. Montillo, R.: The gruesome, true inspiration behind 'frankenstein' | HuffPost. The Huffington Post (2013). https://www.huffingtonpost.com/roseanne-montillo/the-gruesome-true-inspira_b_2622633.html. Accessed 30th Dec 2018
12. Mills, A.A., Day, S., Parkes, S.: Mechanics of the sandglass. Eur. J. Phys. **17**, 97–109 (1996)
13. Bak, P., Tang, C., Wiesenfeld, K.: Self-organized criticality. Phys. Rev. A **38**, 364–374 (1988)
14. Bak, P.: How nature works. The Science of Self-Organized Criticality, Copernicus (1996)
15. Pareto, V.: La courbe de la répartition de la richesse. Œuvres complètes, tome III, Genève, 1967 **III** (1896)
16. Prats, J.: Harry Potter and Pareto's fat tail. Significance (2011). https://www.statslife.org.uk/the-statistics-dictionary/2038-harry-potter-and-pareto-s-fat-tail. Accessed 2nd Feb 2019
17. Gutenberg, B., Richter, C.F.: Seismicity of the Earth and Associated Phenomena. Princeton University Press (1954)
18. Gao, J., Barzel, B., Barabási, A.-L.: Universal resilience patterns in complex networks. Nature **530**, 307–312 (2016)
19. Bloomfield, G., Skelton, J., Ivens, A., Tanaka, Y., Kay, R.R.: Sex determination in the social amoeba Dictyostelium discoideum. Science **330**, 1533–1536 (2010)

20. Brock, D.A., Douglas, T.E., Queller, D.C., Strassmann, J.E.: Primitive agriculture in a social amoeba. Nature **469**, 393–396 (2011)
21. Solow, R.: Technical change and the aggregate production function. Q. J. Econ. **70**, 65–94 (1956)
22. Todar, K.: Growth of bacterial populations. Todar's Online Textbook of Bacteriology. http://textbookofbacteriology.net/growth_3.html. Accessed 26th Dec 2018
23. Science. Amoeba Scale. Time Magazine (1954). http://content.time.com/time/magazine/article/0,9171,890999,00.html. Last accessed Aug 31, 2019
24. Bardi, U.: New research determines the ultimate limits of renewable energy: the bardi sphere. Cassandra's Legacy (2019). https://cassandralegacy.blogspot.com/2019/04/new-research-determines-ultimate-limits.html. Accessed 5th May 2019
25. Berman, A.: The miracle of shale gas & tight oil is easy money Part I—art Berman. Art Berman Web Site (2016). http://www.artberman.com/the-miracle-of-shale-gas-tight-oil-is-easy-money-part-i/. Accessed 23rd Aug 2018
26. Sondland, G.: The fight for EU energy security_2. Politico.eu (2019). https://cassandralegacy.blogspot.com/2019/07/a-rare-glimpse-of-how-elite-think-what.html. Last accessed Aug 31, 2019
27. Malthus, T.: An essay on the principle of population: or, A view of its past and present effects on human happiness. J. Johnson, London (1798)
28. Zubrin, R.: Merchants of Despair. Encounter Books (2013)
29. Ricardo, D.: *The Works and Correspondence of David Ricardo.* (Liberty Fund, 2005)
30. Marchetti, C.: Logistic curves in world history: marchetti and gell-mann. Cambridge Forecast Group Blog (2008). https://cambridgeforecast.wordpress.com/2008/06/21/logistic-curves-in-world-history-marchetti-and-gell-mann/. (Accessed: 1st March 2019)
31. Hubbert, M.: Nuclear energy and the fossil fuels. In: Spring Meeting of the Southern District, American Petroleum Institute, Plaza Hotel, San Antonio, Texas (1956)
32. Bardi, U.: Peak oil, 20 years later: failed prediction or useful insight? Energy Res. Soc. Sci. **48**, 257–261 (2019)
33. Scott Baker, C., Clapham, P.J.: Modelling the past and future of whales and whaling. Trends Ecol. Evol. **19**, 365–371 (2004)
34. Meyer-Gutbrod, E.L., Greene, C.H.: Uncertain recovery of the North Atlantic right whale in a changing ocean. Glob. Chang. Biol. **24**, 455–464 (2018)
35. Laherrère, J.: The Hubbert curve: its strenghts and weaknesses. Oil Gas J. (2000)
36. Guseo, R.: Worldwide cheap and heavy oil productions: a long-term energy model. Energy Policy **39**, 5572–5577 (2011)
37. Bardi, U.: The mineral economy: a model for the shape of oil production curves. Energy Policy **33**, 53–61 (2005)
38. Campbell, C.J., Laherrere, J.F.: The end of cheap oil. Sci. Am. 80–86 (1998)

39. Sweeney, L.B., Sterman, J.D.: Bathtub dynamics: initial results of a systems thinking inventory. Syst. Dyn. Rev. **16**, 249–286 (2000)

40. Perissi, I., Bardi, U., Asmar, T.El., Lavacchi, A.: Dynamic patterns of overexploitation in fisheries (2016). http://arxiv.org/abs/1610.03653. Accessed 26th Oct 2016

41. Perissi, I., Lavacchi, A., Bardi, U., El Asmar, T.: Dynamic patterns of overexploitation in fisheries. Ecol. Modell. **359** (2017)

42. Taagepera, R.: Size and duration of empires: growth-decline curves, 600 B.C.–600 A.D. Soc. Sci. Hist. **3** 115 (1979)

43. McConnell, J.R., et al.: Lead pollution recorded in Greenland ice indicates European emissions tracked plagues, wars, and imperial expansion during antiquity. Proc. Natl. Acad. Sci. USA **115**, 5726–5731 (2018)

44. Bardi, U.: The seneca effect: why decline is faster than growth. Cassandra's Legacy (2011). https://cassandralegacy.blogspot.com/2011/08/seneca-effect-origins-of-collapse.html. Accessed 7th Feb 2019

45. Meadows, D.: Thinking in systems: A primer, Chelsea Greeen (2008)

46. Roddier, F.: Le phénomène de condensation des richesses. Point de vue d'un astronome (2019). http://www.francois-roddier.fr/?p=945. Accessed 5th June 2019

47. Bardi, U.: Why have newspapers become so bad? there is a reason: it is another case of the 'Seneca effect'. Cassandra's Legacy (2015). https://cassandralegacy.blogspot.com/2015/02/the-seneca-cliff-of-newspaper.html. Accessed 9th May 2019

48. Gordon, J.E.: Structures or Why Things Don't Fall Down. Da Capo Press (1978)

49. Meadows, D.H., Meadows, D.L., Randers, J., Bherens III, W.: The Limits to Growth. Universe Books (1972)

50. Redazione.: Tourist killed by falling stone in Florence basilica. ANSA (2017). http://www.ansa.it/english/news/2017/10/19/tourist-killed-by-falling-stone-in-florence-basilica-5_c6ec4fcd-5c39-4c13-ba67-8b949279c8d1.html. Accessed 19th Apr 2019

51. Graves, D.: Field artillery of the War of 1812: equipment, organization, tactics and effectiveness. The Napoleon Series, War 1812 Mag (2009). https://www.napoleon-series.org/military/Warof1812/2009/Issue12/c_Artillery.html. Last accessed 1st Sep 2019

52. ATLAS—the ATLAS project. ATLAS project (2018). http://atlas.fallingstar.com/home.php. Accessed 31st Dec 2018

53. Taleb, N.: The Black Swan. Random House (2007)

54. Sornette, D., Dragon-Kings, Black Swans and the Prediction of Crises. arXiv preprint arXiv:0907.4290v1 [physics.data-an] 18 (2009)

55. Laherrere, J., Sornette, D.: Stretched exponential distributions in nature and economy:"fat tails" with characteristic scales. Eur. Phys. J. B-Condens. **B2**, 525–539 (1998)

56. Kanamaru, T.: Duffing oscillator. Scholarpedia **3**, 6327 (2008)

57. Lotka, A., J. : Elements of Physical Biology. Williams and Wilkins Company (1925). https://doi.org/10.2105/ajph.15.9.812-b
58. Volterra, V.: Fluctuations in the abundance of a species considered mathematically. Nature **118**, 558–560 (1926)
59. Hall, C.A.S.S.: An assessment of several of the historically most influential theoretical models used in ecology and of the data provided in their support. Ecol. Modell. **43**, 5–31 (1988)
60. Serrouya, R., McLellan, B.N., Boutin, S.: Testing predator-prey theory using broad-scale manipulations and independent validation. J. Anim. Ecol. **84**, 1600–1609 (2015)
61. Bernard, L., Gevorkyan, A., Palley, T., Semmler, W.: Long-wave economic cycles: the contributions of Kondratieff, Kuznets, Schumpeter, Kalecki, Goodwin, Kaldor, and Minsky. Soc. Stud. Alm. (2014)
62. Schumpeter, J.A., Opie, R.: The theory of economic development; an inquiry into profits, capital, credit, interest, and the business cycle. Harvard University Press (1934)
63. Clark, C.W., Munro, G.R.: The economics of fishing and modern capital theory: a simplified approach. J. Environ. Econ. Manag. **2**, 92–106 (1975)

3

The Practice of Collapse

I'm truly sorry man's dominion,
Has broken nature's social union,
An' justifies that ill opinion,
Which makes thee startle
At me, thy poor, earth-born companion,
An' fellow-mortal!
Robert Burns: "To a Mouse"—1785

The Collapse of Engineered Structures: Dust Thou Art, and unto Dust Shalt Thou Return

Fig. 3.1 A balcony photographed in 2019 near Florence, Italy. You can see the badly corroded iron beams appearing through the cracks of the concrete structure. This balcony is in a dangerous condition and it might collapse under stress but, apparently, the owners of the building cannot afford to have it repaired, a common condition for many reinforced concrete structures, all over the world. The pigeon, of course, does not care! (photo by the author, 2019)

© Springer Nature Switzerland AG 2020
U. Bardi, *Before the Collapse*,
https://doi.org/10.1007/978-3-030-29038-2_3

In the late morning of August 14, 2018, I was busy writing this book when I happened to open my browser. There, I saw the images of the collapse of the Morandi bridge, in Genoa, almost in real time. It was a major disaster: the bridge used to carry more than 25 million vehicles per year and it was a vital commercial link between Italy and Southern France. When it collapsed, it not only took with it the lives of 43 people who were crossing it, but it was nothing less than a stroke for the Italian highway system, forcing the traffic from and to France to take a long detour. It will take years before a new bridge can be built and the economic damage has been incalculable (Fig. 3.2).

Fig. 3.2 The remains of the Morandi Bridge (or Polcevera Viaduct) in Genoa, Italy, after a whole section collapsed on August 14, 2018. (Image by Michele Ferraris, Creative Commons)

How could it be that the engineers who took care of the maintenance of the highway could not predict and contrast the collapse of such an important structure? Much was said in the debate that followed about incompetence or corruption. Perhaps the fact that maintenance of the highway was handed over to a profit-making company was a recipe for disaster: profit-maximizing may well have led to cutting corners in the maintenance tasks. But, on the whole, we have no proof that the company that managed the bridge was guilty of criminal negligence. Rather, the collapse of the Morandi bridge may be seen as another example of how the behavior of complex systems tends to take people by surprise.

Even in engineering, with all its emphasis on quantification, measurements, models, and knowledge, the phenomenon we call "collapse" or "fracture" remains something not completely mastered. If engineers knew exactly how to deal with fractures, nothing ever would break—but, unfortunately, a lot of things do, as we all know. We saw in a previous section how critical phenomena in a network can be initiated by small defects in the structure, it is the effect of *cracks* in real-world structures, according to the theory developed by Alan Griffith [1]. The Morandi Bridge was a structure under tensile stress, sensible to the deadly mechanism of the Griffith failure.

The bridge went down during a heavy thunderstorm and that may have been the trigger that started the cascade of failures that doomed the bridge: one more case of the "Dynamic Crunch" phenomenon that leads to the Seneca Cliff. Somewhere, in one of the cables holding the deck, there had to be a weak point, a crack. Then, perhaps as an effect of a thunderbolt, or maybe of the wind, the cable snapped off. At that point, the other cables were suddenly under enhanced stress, and that generated a cascade of cable failures which, eventually caused a whole section of the bridge to crash down. You heard of the straw that broke the camel's back, in this case we could speak of the lightning bolt that broke the bridge's span. Complex systems not only often surprise you. Sometimes, they kill you.

But why was the Morandi bridge so weakened? Just like many other bridges in Italy and Europe, it had been built using "pre-compressed concrete." This is a material European engineers seem to like much more than their American colleagues who, on the contrary, tend to use naked steel cables and beams for their bridges. Pre-compressed concrete had more success in Europe because it was widely believed that concrete would protect the internal steel beams from corrosion and avoid the need for laborious maintenance work of painting and repainting required, instead, for steel bridges. But, over the years, it was discovered that steel corrodes even inside concrete, and that turns out to be a gigantic problem, not just for bridges.

In the case of the Morandi bridge in Genoa, the problem was known. The bridge had been opened in 1967 and, after more than 50 years of service, it needed plenty of attention and maintenance. Years before the collapse, engineers had noted that corrosion and the vibration stress caused by heavy traffic, had weakened the steel beams of the specific section that was to go down in 2018. A series of measurements carried out one year before the collapse had indicated that the steel in that section had lost 10% to 20% of its structural integrity. That was not considered to be dangerous enough to require closing the bridge to traffic, especially at the height of the busy summer season. After all, most buildings are built with a hefty safety margin with respect to their breakdown limit, typically at least 100%. But there was a plan to close the bridge for maintenance work in October 2018. Too late.

We see once more how the best plans of mice and men often go astray. The engineers who were working on the bridge may have made a typical mistake of linear thinking: they assumed that there is a certain proportionality between weakening and danger. In this case, they believed that a 20% weakening of the beams was not enough to cause the bridge to collapse. But that was an average, and complex systems may not care about averages: do you know the story of the statistician who drowned in a river of an average depth of 1.5 meters?

Bridges are just an example of the many engineered structures subject to collapsing under stress. The Griffith mechanism of crack propagation is typical of the fracture of structures under tensile stress, such as the beams of a suspension bridge, the beams of a roof, moving objects such as planes and ships, everyday objects such as bookshelves, and even the bones of living beings. These structures tend to go down rapidly, suddenly, and sometimes explosively, typical examples of Seneca Collapses. There also exists another category of engineered structures, those which must withstand only compression stresses: this is the case of pillars, walls, arcs, domes, and the legs of the chair you are sitting on. These structures can collapse, but are normally much safer than those under tension because compression tends to close cracks instead of enlarging them, as tension does.

In ancient times, when reinforced concrete did not exist, buildings used to be made in such a way to avoid all kinds of tensile stresses as much as possible. That was because the main construction material available in ancient times was stone and stone just cannot take tensile stresses. So, stones can be used to build walls and buttresses, and also for bridges and roofs, provided that you arrange them carefully to form arcs and domes in order to make sure that all the elements are always under compression, never under tension.

But even compression structures have their limits. Ancient builders were perfectly aware that stone can crumble, even explode, when subjected to excessive stress. That generates a limit to the height of a building in stone: over a certain height, the stones at the base would burst out and bring the whole structure down. One of the arts that ancient builders needed to know was the capability of testing stones for their resistance to compression and they had developed sophisticated measurement techniques to determine this property. Maybe we are biased in our perception because what we see around us are only those ancient building which survived and arrived to our times, but it is true that many ancient buildings have survived the test of time beautifully and are still around us after several centuries, even millennia.

Many Roman bridges are still standing and are used today. Another remarkable example of a building that survived from Roman times is the *Pantheon* temple, in Rome. It was built nearly 2,000 years ago and it still being used as a temple today, now a Catholic church. Gothic cathedrals built during the Middle Ages were also sturdy and resilient: there are only few examples of structural collapses caused by poor design. For instance, the Beauvais Cathedral, in France, built mainly during the 13th century, suffered lots of problems and some structural collapses, but it is still standing nowadays. Another example is the Pisa tower, in Italy, built during the 14th century. For centuries, it survived the bending caused by ground movements. During the 20th century, the bending had reached an angle of 5.5°, bringing the tower to risk collapse. Today, the tilt has been reduced to less than 4° by acting on the foundations, and now the tower may well keep standing for more centuries in the future. Modern stone buildings are sometimes even more ambitious. The Washington Monument in Washington DC is an example of a building high enough (169 m) to be close to the limits of structural resistance of the stones at its basis. It was terminated in 1884 and seems to be still in good shape despite some cracks that it developed after an earthquake hit it in 2011.

As a last note on this classification, I could mention the "Euler Collapse," a mode that mixes something of the tensile and something of the compressive elements of the fracture mechanism. It occurs when a thin structure is subjected to compression and, as a consequence, it twists sideways. An example is what may happen to women when they walk on high heels. The tensile stresses at the heel may break it at the juncture with the sole or, in the worst case, fracture the wearer's ankle. Wearing high heels is dangerous, but many ladies seem to like the idea. I may tell you that once I was in a Russian town in winter and I saw a young lady on high heels running to catch a bus over the iced sidewalk, jumping inside gracefully and apparently at ease. How she could do that without slipping on the ice and killing herself, or being run over by the bus, remains a mystery to me to this date. Maybe you have to be Russian to be able to do certain things. But humans are complex systems and complex systems always take you by surprise.

But let us go back to the case of the Morandi bridge for a discussion on risk evaluation about engineered structures. I crossed that bridge by car several times in my life without ever even vaguely thinking that it was risky to do so. Probably, at least a billion vehicles safely crossed that bridge over its more than half a century of life, so the chance of seeing it collapse just when you were crossing it was abysmally low. Yet, it happened in 2018, and when a major bridge collapses someone is bound to be crossing it. Obviously, it would have made no sense to avoid crossing the Morandi bridge, or any other concrete bridge, for fear that it could collapse. Yet, it makes perfect sense to consider the risk of collapse for a building that you use much more often than bridges: your home or the place where you work. Unfortunately, normally you have no idea of how well and carefully your home was built and maintained. Maybe all the standards were respected, maybe not and, in the second case, your life is at risk: the collapse waiting for you could be rapid and deadly.

There are many cases when it was discovered, typically after the collapse of a structure, that the builders had saved money by reducing the amount of steel reinforcement for the concrete. Or maybe they had used poor quality sand; a typical trick to save money is to use sand taken from some beach. This sand is contaminated with sea salt and that favors the corrosion of the steel beams inside the concrete. In some cases, it is reported that instead of the standard steel beams, builders used wire mesh of the kind used for chicken coops. Then, you have to consider that a building rarely remains untouched after it has been built. People open doors and windows in the walls, add more floors, remove walls or add them. They may also intervene in other dangerous ways: for instance, everyone loves rooftop swimming pools, but they are heavy and may destabilize the whole structure of a building. These mongrel buildings may be very dangerous: one of the worse disasters in the history of architecture happened to a building that was modified and expanded without much respect for rules or for common sense. It is the case of the *Rana Plaza* collapse on April 24th, 2013 in Savar, a district of Bangladesh, when more than one thousand people died and more than 2,500 were injured. The owners had added four floors to the building without a permit (!!) and also placed the heavy machinery of a garment factory in these extra floors. Not only was he machinery heavy, but it also generated strong vibrations that further weakened the building. More than half of the victims were women workers of the factory, along with a number of their children who were in nursery facilities within the building. A good example of criminal negligence.

Building collapses are rare, so the risk is so small that it is not normally listed in the various "Odds of Dying" tables that you can find on the Web [2]. Yet, it is one of those risks for which you can take precautions and there is no reason for not doing so. If you live in a building made of reinforced concrete that is older than a couple of decades, you should check for the details that may indicate danger. In some cases, you can directly see the corrosion of the steel beams where the surrounding concrete has been eroded. Cracks in the walls are an evident symptom of troubles and it has been reported that the noise of a steel cable snapping open inside a concrete beam may be perceived as the noise of gunshots. In Europe, if you hear that kind of noise, you may reasonably think that there is something wrong with the structural integrity of the building you live in, but, of course, gunshots may be much more likely if you live in the US. By the way, the collapse of the Morandi bridge gave rise to noises that could be interpreted as explosions and—guess what!—that led some people to interpret the disaster as the result of a "controlled demolition" carried out by the evil "Zionist Illuminati" in analogy with the demolition theories proposed for the 2001 attack to the world trade center in New York [3]. Human fantasy seems to have no limits in terms of crackpot theories.

Not seeing or hearing anything suspicious in a building does not necessarily mean it is safe. If it is older than 50 years, it would not be a bad idea to seek professional help to have it checked for its structural integrity. It is expensive, though, and not routinely done for private buildings. Stone buildings are normally safer and more durable than concrete ones; you have to be careful, though, because these buildings can crumble under the effect of lateral vibrations generated by earthquakes. Wooden houses are often said to be more resilient and safer than both concrete and stone buildings and that is probably true, within some limits. But take into account that wooden beams are susceptible to degradation, too: they may be attacked by termites and their presence may be difficult to detect because they eat away the interior of the wood before breaking through to the surface. In terms of structural safety, an Indian tepee or a Mongolian yurt would be the best choice for a place to live. Otherwise, you have just to accept that there are some risks in life.

In the end, the problem of concrete degradation is not with single buildings: it is a global problem that affects all the infrastructure built over the past century or so (Fig. 3.3).

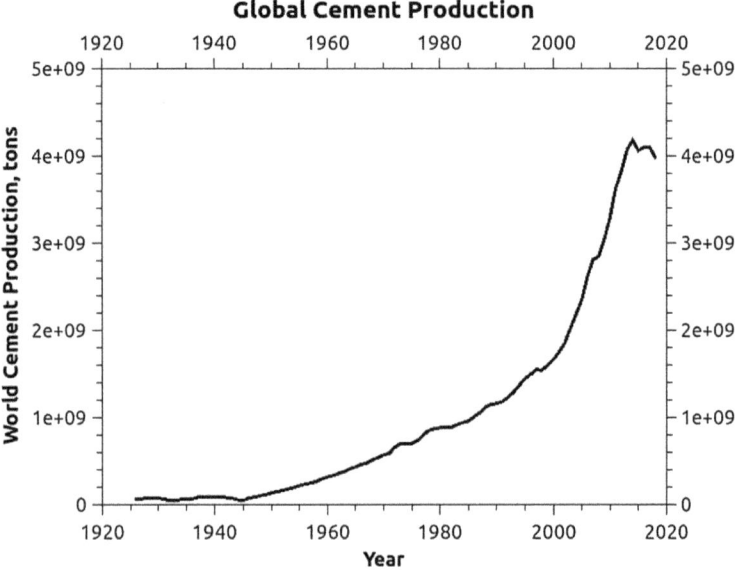

Fig. 3.3 Global cement production. Data from USGS

You see in the figure how concrete production went through a burst of exponential growth from the 1920s all the way to a few years ago. Only in 2015 did the global production of concrete start to show signs of stabilizing and, probably, it will go down in the coming years. It means that our highways and our cities were built in a period of economic expansion and on the assumption that the needs for their maintenance would have been minimal, just as it had been for the previous generation of stone buildings. It turned out to be a wrong estimate.

In the future, we seriously risk an epidemics of infrastructure collapses if we do not allocate sufficient resources to the maintenance of their concrete elements. Otherwise, the result could be that a considerable fraction of the world's buildings and roads will have to be sealed off and left to crumble. Worse, crossing a bridge or living in a skyscraper could come to be considered risky. It is already the situation you have in some poor countries. In Cuba, after the revolution of 1959, the government expropriated most buildings that had been owned by rich Cubans and foreigners and distributed them among the poor. The problem is that these buildings had been erected using Portland cement made from beach sand contaminated with sea salt. Sea salt favors the corrosion of the steel beams—it is a very serious problem. It can be remedied, but it is expensive and requires sophisticated technologies [4] that Cubans cannot afford today. The problems of old concrete buildings in poor countries do not seem to be related to a specific political ideology or government system. Puerto Rico is under the control of the American government but the problem of crumbling buildings seems to be the same as in Cuba [5], worsened in recent times by the Hurricane Maria that struck the island in 2017 [6]. Other areas with warm climates and close to the sea seem to be affected in the same way.

We lack worldwide statistical data for this kind of problems, but there seems to exist a "crumbling belt" of decaying buildings everywhere in tropical regions, especially near the sea, where higher temperatures and sea salt spread by the wind cause the steel beams of concrete building to corrode faster than in other regions of the world—incidentally, the Morandi Bridge was near the Mediterranean coast and it may well be that in that case too, sea salt had a role in the collapse. Add to that the fact that in many of these regions people are poor and unable to afford the costs involved in the remediation of these old buildings, and you have a big global problem: another Seneca Cliff awaiting.

In the end, the problem has to do with an old Biblical maxim: "dust thou art, and unto dust shalt thou return." Applied to a concrete structure, it would sound more like, "sand thou art, and unto sand shalt thou return." Concrete is nothing else than compacted sand, not unlike the sandcastles that children build on the beach. The substance that binds the sand in sandcastles is water, and when it evaporates the castle crumbles. In concrete, the binder is cement, and it is typically lime or calcium silicate. Of course, this kind of solid binder doesn't evaporate and concrete lasts much longer than sandcastles, but not forever. So, what we are seeing today in Cuba and other poor tropical countries may be just an image of what our world will be in a not-so-remote future.

The risk of a collapse affects all kinds of engineered structures, not just buildings. Among the countless objects that humans build, many are especially dangerous because they move—sometimes very fast. According to the available statistics [2], pedestrians are the most likely victims of street accidents while the most dangerous kind of vehicles are motorcycles. The odds of being killed in a car accident in the US are about 1 in 10,000 every year, a value that we do not consider as worth worrying about because most of us normally use cars and walk in streets where the risk of being hit by a car exists. Planes are significantly less dangerous than cars. According to The

Economists [7], a typical value for the probability of being killed in a plane crash is one in 5 million per flight. Even if you were to take a flight every day for a year, the chance of being killed in a crash would be less than one in 10,000, not really worth worrying about.

Although these odds are small, they are not negligible and most of us have relatives or friends who suffered a major road transportation accident. The question is how to reduce the chances of being involved in one. In the case of road transportation, there are many well known rules and recommendations about the things to do and not to do when putting oneself at the wheel. But when you ride a vehicle driven by someone else, for instance when you take a bus, you have no idea of the competence of the person at the wheel: the driver may be incompetent, drunk, under the effect of heavy drugs, or, worse, harboring suicidal thoughts. On this point, it may be worth remembering the recommendations made by Jared Diamond in his book *The World Until Yesterday* (2013) where he tells us how he nearly drowned when a small boat carrying him and a few others was sunk by a reckless New Guinean pilot. Diamond notes that he should have noticed that there were problems with that boat before boarding it if he practiced the art that he noted in his New Guinean friends and that he calls "constructive paranoia." It is a set of habits involving extreme attention to details of potentially dangerous people and objects, developed by people who live in more challenging environments than those typical of our experience of Westerners. Overall, though, you cannot use paranoia as the way to manage your life. You have to accept that perfect safety is something that you can have only inside your grave.

Nevertheless, you may improve your chances of surviving by exercising a certain critical attitude with choosing your transportation system. There is much discussion on whether some airlines are safer than others, but a comparison is often difficult because there are many factors involved - the route, the kind of planes, the number of flights but, more than that, because the number of disasters in the airline industry is so small (fortunately) that a statistically significant comparison is nearly impossible. It is also true that not all planes are the same and you might think you could choose a flight on the basis of which model of plane will be used. But that is rarely specified in the ticket and can be changed anyway according to the needs of the airline. When you buy an airline ticket you automatically agree to the contract called "conditions of carriage" which is normally a ponderous document that nobody ever reads. In the US, every airline has a different contract, but they tend to be very similar. For instance, the conditions of carriage of Delta airlines in 2017 specified that [8] "Delta may substitute alternate carriers or aircraft, delay or cancel flights, change seat assignments and alter or omit stopping places shown on the ticket at any time. Schedules are subject to change without notice." And notice that they do not even say that they will take you there by a plane—they only mention "alternate carriers" which might be a camel caravan. Fortunately, that does not happen very often.

So, you have no way to know what kind of plane the company you chose will use, nor whether it will be a new plane or an old one, and whether it could have had maintenance problems in the past. For instance, the people who boarded the Aloha Airlines flight 243 en route from Hilo to Honolulu, in 1988, had no way of knowing that the plane—an old version of the Boeing 737—had a serious problem. Having been employed for several years on that route, it had undergone a much larger number of cycles of compression and decompression than similar planes employed on longer hauls. These numerous cycles had weakened the hull and, as a result, the plane lost part of the fuselage in mid-flight. It was another case of critical failure generated by the mechanism of the expansion of a Griffith crack. In that case, fortunately, the pilots managed to land the damaged plane in Honolulu, still in one piece, although minus a big chunk of the fuselage. The pictures taken after the landing show the passengers still sitting in their seats in the open, it was as if the plane had been turned into a convertible car. We can only vaguely imagine how these people must have felt finding themselves, literally, sitting in the middle of the sky when the plane opened up like a tin can. Sadly, one of the flight assistants died when she was sucked out of the plane, but the survival of the other passengers and crew was nothing less than a small miracle.

Occasionally, though, you do have a choice for which plane to board. You surely heard of the recent case of the crashes of two Boeing 737 "Max" planes in 2019 caused, probably, by a faulty design of the control software [9]. Several state regulatory agencies all over the world grounded all the 737 Max planes immediately, but in the US, the plane continued to fly for some days. In that case, you had a choice on whether to fly with an airline that still used the Boeing 737 or not. You might have been paranoid enough to choose a European or a Chinese airline instead of an American one.

In the end, when we travel we tend to lock ourselves up inside metal boxes running on roads or flying in the sky at speeds such that crashes will often be fatal. Statistically, someone will have to be hit by this specific kind of Seneca cliff. This, too, is part of the rules of the universe.

Financial Collapses: Blockbuster Goes Bust

A Scene from Shakespeare's 'Merchant of Venice', Act 4, Scene 1: Shylock and Portia, 1790, David Allan

Fig. 3.4 Illustration from Act 4 of Shakespeare's The Merchant of Venice. The evil money-lender Shylock demands one pound of flesh from Antonio because he is unable to repay his debt, an illustration of how harsh the penalties for insolvency were in ancient times

Imagine that the year is 2008 and that you are the CEO of Blockbuster: a large, international company specialized in movie rentals. At its peak, in 2004, Blockbuster was employing more than 80 thousand people with almost 10,000 retail stores worldwide and a global yearly revenue of some 6 billion dollars. But, in the following years, the company stopped growing. As CEO, you realize that there are problems, but it is also true that Blockbuster remains the top dog in the market. True, you have a competitor: a newcomer called "Netflix," they are aggressive and they are growing. They have even proposed to you a merger, but why should you accept to merge with a smaller company? There is no reason for Blockbuster to make big changes, the current slowdown is just a temporary downturn, it can surely be remedied by trimming expenses and improving efficiency. Then, in 2009, Blockbuster suddenly loses more than 20% of its revenues. One year later, the company is bust and you are out of your job (Figs. 3.4 and 3.5).

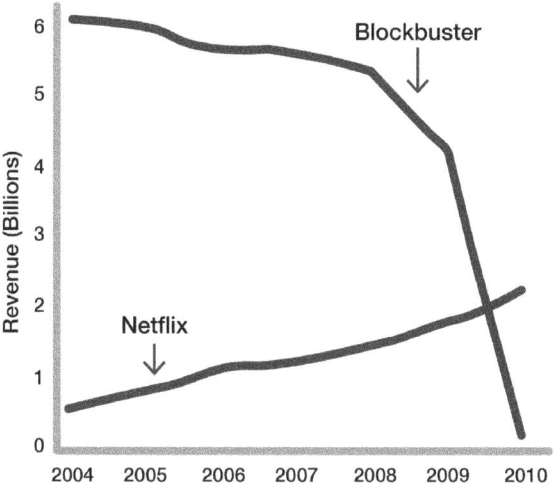

Fig. 3.5 The collapse of Blockbuster and the rise of Netflix. Data from [10]

Why did Blockbuster go down so fast? Mainly because the company was caught with a marketing strategy that had become obsolete. Greg Satell reports on "Forbes" [10]

> Blockbuster's model had a weakness that wasn't clear at the time. It earned an enormous amount of money by charging its customers late fees, which had become an important part of Blockbuster's revenue model. The ugly truth—and the company's Achilles heel—was that the company's profits were highly dependent on penalizing its patrons.

Netflix had a different approach: ordering was done online, monthly subscriptions were flat-rate, and there were no late fees. Then, Netflix was a pioneer in offering online streaming services, Blockbuster followed, but too late and with a less effective plan.

When things started going bad for Blockbuster, we can imagine the bells ringing in the company headquarters. Surely there were meetings of the managers desperately trying to "do something" to stave off the disaster. And, just as surely, a lot of "solutions" were devised and some put into practice. But it was too late: the management of Blockbuster was taken by surprise and the deadly mechanism of enhancing feedbacks had kicked in: the more Blockbuster was losing money, the more the debt it accumulated made it difficult for Blockbuster to propose good deals to its customers. And with customers leaving Blockbuster, more money was lost and more debt accumulated. Until the bitter end, in 2010.

Hemingway probably never heard of the concept of Seneca Collapse, but he described it perfectly well in his 1926 book, *The Sun Also Rises*:

"How did you go bankrupt?" Bill asked.
"Two ways," Mike said. "Gradually and then suddenly."

Debt, in itself, need not be a bad thing: you may argue that without debt the whole society could not function. David Graeber provided a general history of debt in his book "Debt, the first 5000 years" [11] showing how debt is the very essence of money, something that had been argued first by Mitchell Innes in 1914 [12]. But the problem with debt is that it accumulates, often beyond the practical possibility to repay it. Then, debt becomes insolvency and that's bad. In all human societies, not being able to maintain one's promise is a serious breach of trust, something that may destroy the very fabric of a family, a company, or an entire state. Monetary insolvency is just a quantified version of breaking a promise.

In ancient times, people unable to repay their debt faced harsh laws and customs that we could term as "draconian." The early Roman laws were based on the concept of *manus iniectio* (literally, "hand lain on him") which could mean that the insolvent debtor could be physically punished, perhaps killed or reduced into slavery. A remnant of these ancient laws can be found in Shakespeare's *The Merchant of Venice* when the moneylender Shylock insists in taking a pound of flesh from the protagonist, Antonio, when the latter cannot pay back his debt. Surely a bad case of a Seneca Collapse for him.

So harsh were the penalties for insolvency that most ancient law codes also included provisions for leniency. Already in Sumerian times, as early as three millennia ago, there existed a use called Ama-gi (or Amar-gi) [13], a term translated as "freedom" but, literally, "return to the mother" that involved the periodic wiping out of all debts. The Jews had similar traditions with the *Shemitah* (Sabbatical) and the Jubilee (every seven sabbatical years), when various obligations were forfeited, including debt. The periodic cleaning ups had the function of avoiding excessive accumulation of debt.

The Jubilee was a good idea, but it carried a big problem: when the year of the cleanup was approaching, then nobody would loan anything to anyone knowing that soon his credit would be canceled. That was probably the reason why Rabbi Hillel the Elder introduced the *prozbul* rule during the 1st century BCE. It allowed the stipulation of contracts with an explicit clause that would make the debt immune from the periodic wiping out of the *Shemitah*.

Other kinds of legislation did not involve the erasure of all debt, but reduced the penalties on the insolvent debtor. Already during Roman Imperial times, the punishment for insolvency was considerably softened in comparison with earlier times. During the Middle Ages in Europe, the penalty for insolvency could be limited to a public humiliation to be carried out on a special stone called *lapis scandali* [14] or the "stone of scandal." The debtor was forced to sit, naked, on the stone while he had to claim loudly that he forfeited all his goods to the creditor. That was humiliating, surely, but not as bad as having to give a pound of one's own flesh to the creditor. The idea of softening the punishment for insolvency cuts across many societies and cultures and, in the Koran, in the *Sura al-Baqarah* (Sura of the Cow), we can read at verse 280: "And if someone is in hardship, then let there be postponement until a time of ease. But if you give from your right as charity, then it is better for you, if you only knew." In modern times, bankruptcy laws are varied and depend on the specific legal systems of different countries. The general idea, anyway, is always the same: to soften the impact of insolvency on both the debtor and the creditor.

These current bankruptcy laws are surely not perfect, but they are badly needed since financial collapse is a very common event. According to Eric Wagner, writing on Forbes [15], 80% of startup companies fail within the first 18 months. Bankruptcy is normally imposed by court order. The court nominates a bankruptcy trustee who then liquidates the assets of the insolvent company or person and distributes the proceeds to the debtor's creditors. Then, the bankrupted person or company can restart anew, at least in theory. In practice, it is not always possible to wipe out one's debt so easily. For instance, in the United States, you may have obtained a federal student loan on the basis of the idea that the rates are low, but you may still find that you cannot pay your debt back. In this case, you may discover that the option of personal bankruptcyis not available to you, except in special cases. You are indebted for life in a condition that starts looking like a form of slavery [16]. Much worse than having to sit naked on a stone for a while.

Even when it works the way it is designed to do, bankruptcy may have bad consequences on everyone involved. Small scale bankruptcies may generate the foreclosure of one's house, which is a serious trauma that may affect people for the rest of their life. Debt and bankruptcy can result in symptoms of PTSD (post-traumatic stress disorder) leading to depression and even suicide [17]. Large scale bankruptcies involve the loss of jobs for thousands of people and, at very large scales, the result may be political instability, civil wars, and more disasters. Clearly, financial collapses are a bad kind of Seneca collapse and, as a first approach, the problem boils down to avoiding to get caught in one. There is little that you can do if you live in a country which goes down in a major financial collapse: all you can do is to try to survive the best you can. But, on a smaller scale, it is possible to take precautions.

There are many recipes you can find in books and internet sites on how to invest your savings in such a way to multiply them by a large factor and make you rich. An incredible number of these recipes are evident Ponzi schemes designed to siphon your money out. Just as an example among the many, I can cite the scheme called "Quantum Code" [18]. At the moment I am writing this chapter, early 2019, the web site of the Quantum Code company still exists and you can find it by googling various combinations of "quantum code," "financial," and "investing." The clip they use to peddle their scheme is very well done and, over and over, you are shown all the perks of being rich: a personal jet plane, big cars, jewels, expensive trinkets, and more. Art, after all, is mostly based on some kind of make-believe process and when we watch a play by Shakespeare we do not worry about whether Hamlet is a historical character or not. In this case, the actors playing the alleged financial tycoon "Michael Crawford" and his personal assistant, "Tasha", do a superb acting job.

It is clear that the scheme is a scam from the first sentence you hear in the clip: "*my name is Michael Crawford, yes, that guy you might have read about on Forbes and other financial magazines.*" It takes less than one minute to verify that there does not exist anyone with that name mentioned on Forbes or any other magazine as a financial tycoon or anything like that. Maybe a lot of people, out there, are unable to use search engines for debunking this kind of stories. Still, anyone should be wary when hearing "Michael Crawford" telling them that he wants to make them millionaires in exchange for nothing, out of pure philanthropy. Don't they have a grandmother who told them that "there ain't no such thing as a free lunch"? So, how would anyone believe in this so transparent scam?

But for everything that exists, there has to be a reason for it to exist. The fact that the Quantum Code clip is so easily debunked cannot be a bug; it must be a feature. The scam is so transparent that we can only imagine that the script of the clip was thought from the beginning as a sucker's bait.

Evidently, they *want* suckers and they make sure that those who fall for the trap *are* suckers. Indeed, it can be shown that the trick of pre-selecting suckers is the best strategy in order to optimize the effort of the scammer. We can read a discussion of this point in a paper by Herley [19]. (Here, "attack" refers to the decision of the scammer to engage with the target).

> The endgame of many attacks require per-target effort. Thus when cost is non-zero each potential target represents an investment decision to an attacker. He invests effort in the hopes of payoff, but this decision is never flawless.

The general idea is that scammers tend to pre-select targets stupid enough to fall into a transparent form of scam and are, therefore, nearly sure victims of the scam. After all, it is nothing different from the strategy that lions and leopards tend to use when they choose their prey among the weak and among isolated individuals.

Of course, different people need to be cheated differently. A cursory examination of the World Wide Web shows that there exists a whole market of financial scams "graded" for different customers. An extreme case of obvious scams is that of the "Nigerian Scam," sometimes known as the 'Nigerian 419' scam, since the first wave of this scam came from Nigeria and '419' comes from the section of Nigeria's Criminal Code which (in theory) outlaws it. It works like this: the victim will receive a message telling an elaborate story about large amounts of money trapped in a bank during events such as civil wars or coups, or maybe because of an inheritance blocked by government restrictions or taxes. The scammer will then offer the victim a large sum of money to help them transfer that money out of the country.

I don't think anyone among the readers of this book needs to be warned not to fall into a scam so obvious as the Nigerian 419 but, as I said, there is a whole zoo of scams at various levels. The Quantum Code is an easily detectable one, although it is much more sophisticated than the Nigerian 419. Climbing up the ladder, we find theoretically serious schemes such as the various kinds of "hedge funds." The idea of these funds is to use sophisticated risk management techniques in order to diversify investments and reduce the risks for investors—today, hedge funds manage several trillion dollars worldwide. However, it is debatable that these funds can really protect investors from systemic risks such as a global market collapse, as happened in 2008. In addition, according to Nassim Taleb [20], the hedge funds are vulnerable to the "black swan" collapse, in this sense they have some point in common with the "martingale strategy" at the roulette—doubling the bet after each loss. Just as for the martingale strategy played at the roulette table, hedge funds may only trade risks: they may reduce the frequency of small losses in exchange for a low frequency, but not impossible—large loss.

A good example, here, is the case of Amaranth Advisors. We can read in "Investopedia" [21] that

> After attracting $9 billion worth of assets under management, the hedge fund's energy trading strategy failed as it lost over $6 billion on natural gas futures in 2006. Faced with faulty risk models and weak natural gas prices due to mild winter conditions and a meek hurricane season, gas prices did not rebound to the required level to generate profits for the firm, and $5 billion dollars were lost within a single week. Following an intensive investigation by the Commodity Futures Trading Commission, Amaranth was charged with the attempted manipulation of natural gas futures prices.

In this field, we all have to be very careful because none of us is immune from the Dunning-Kruger trap [22]. It is a syndrome that makes people think they are smarter and more knowledgeable than they really are. And no matter how smart you are, there is a probably a scam exactly tailored for you, somewhere. I could tell you stories about my own experience, even though, fortunately, never involving major financial losses. But every time I hated myself for having been so naive to fall for such obvious tricks—and yet I did. But so is life, they say that a sucker is born every minute and every one of us can be a sucker in some circumstances.

There is one more case of financial Ponzi scheme worth a note, here: "technological scams," a field in which I can claim a certain degree of experience because of my job as scientific researcher. This kind of scams is based on the diffuse idea that technology can produce miracles. That, in turn, is based on the fact that during the 20th century we saw the development of new technologies that could be described as nearly miraculous: think of antibiotics, nuclear power, electronic devices, and much more. But that does not mean technological miracles can be obtained at will. There remains valid the basic rule that says that progress is based on "1% inspiration and 99% perspiration." Professionals in innovation know this, but ordinary people often do not and their naive faith in advanced technology may make them easy victims of technological scams.

There exist plenty of people and companies claiming to possess wonderful technologies able to solve this or that world problem. Some of these ideas are serious ones, proposed by serious people, that deserve attention for future developments. But many are over-hyped and in not a few cases they are outright scams. The fauna in this area involves a variety of types, from the solitary mad scientist to well-intentioned but misguided efforts destined to fail because of the realities of physics or of the market.

In the category of the "mad scientists," a mention should surely go to the Italian inventor Andrea Rossi, known for his "energy catalyzer" or "E-Cat," a device supposed to produce energy by the nuclear fusion of hydrogen (or perhaps of some other element, or perhaps none at all) [23]. Surely, Rossi has a certain knack in promoting himself and his ideas. He succeeded in peddling his E-Cat to the Department of Physics of the University of Bologna [24] making the members of this ancient and respected institution suffer a considerable loss of prestige. Using his association with the university as a certificate of seriousness, Rossi's invention went through the Web as a bright meteor that for a while even reached the mainstream media. Today (2019) the E-Cat seems to have lost interest and faded away, even though Mr. Rossi is still active in promoting it.

Rossi's scheme is a typical example of many similar ones I have seen in my career. It goes like this: someone shows up at the door of a department of a university or of a research institution. The person proposes a hefty grant to the researchers to test and improve the wonderful process he or his company are developing. If the university or the institute accepts the grant, the money involved may or may not be paid, but the inventor(s) will use the grant to claim that the idea has been validated by the university or the research instutute. Rossi had promised Eur one million to the University of Bologna, which he never paid. It is reported that he tried to play the same trick with NASA [25].

Something similar happened to me. Years ago, someone asked the University of Florence to test a new method to produce ultra-pure silicon that his company had been developing. It looked like a serious proposal and the physics involved was sound. So, we accepted the grant and two researchers of my group worked on that subject for about a year. We found that the process worked, at least at the laboratory scale, and that it surely deserved more efforts to be developed at the industrial level. But we soon discovered that the proposers had no intention of exploiting the new process, they only wanted cash from the government and some big investors. And they did not even want to pay us. Fortunately, the legal office of my university could force them to shell out most of the money they had promised in the contract. Afterward, we never heard of them again.

These stories illustrate how difficult it is to invest in technological schemes: some are scams, many are just bad ideas, and even a good idea can turn into a scam if the people promoting it are financial sharks. As someone said, there are three ways to ruin oneself: the most pleasant one is with women, gambling is the fastest, high technology is the surest.

Returning to the problem of bankruptcy, clearly, insolvency is one of those elements of our society that we would like to ignore but that may badly affect us anytime during our lives. But what is insolvency, actually? And why does it exist? In standard economics, bankruptcy is dealt with in the two main branches of the field. "Microeconomics" studies the behavior of individuals and firms in making decisions in allocating resources and structuring production and other characteristics. "Macroeconomics" takes a larger view, studying the economy as a whole, in particular in terms of the effect of government policy decisions.

Microeconomics uses a variety of models aimed at finding the optimal values of the parameters of a firm or of a process. It may also assume a qualitative aspect when it examines what decisions managers take to steer their companies through the perilous waters of that entity we call "the market." It is the same challenge faced by individuals and families trying to navigate in a difficult world: paying the mortgage for the house, feeding the children, repairing the car, all that. Bankruptcy may ensue because someone makes a wrong decision—as for Blockbuster that of not following the evolution of the market. Or, it may happen because something changes all of a sudden, say, one loses one's job and cannot find a new one. Overall, microeconomics gives us many examples from which to learn, but no general theory of why economic entities collapse.

Macroeconomics, instead, aims at understanding how the economic system works and that includes also financial collapses, obviously part of the system. Here, Hyman Minsky developed the "Financial Instability Hypothesis" [26] starting in the 1970s. I think Minsky's idea can be summarized as "success breeds excess." That is, during periods of economic growth people tend to become excessively optimistic, they borrow heavily from banks, and find themselves in a spiral of debt that soon goes out of control. Then, investors want to be paid back and that generates a cascade of reinforcing feedbacks bringing the whole company crumbling down in a classic Seneca Cliff. It looks very much like the story of Amelia the amoeba that we saw in an earlier chapter: a biological population that grows exponentially until it crashes down when the food runs out. In the case of a company, money plays the role of "food" and uncontrolled growth makes the company run out of the food it needs.

Eventually, the whole problem of financial collapses is the result of the existence of money. But what is money exactly? Without going into the various theories of money that economists are still discussing, we can say that, once, money was something that everybody agreed on: a weight of precious metals. After all, the British currency is still defined in units of weight, even though one pound (in monetary terms) does not weigh a pound (in physical terms). Still, up to not too long ago, money was simply a token representing a physical entity: a certain weight of gold or silver. But things changed a lot with time and, with the 20th century, the convertibility of the dollar into precious metals became more theoretical than real. In 1971 President Nixon formally canceled it. From then on, money has been a purely virtual entity, created by central banks out of thin air. How can it be that people accept to be paid for their work with something that does not exist is a little strange, if you think about that. But that does not change the fact that money is the backbone of society: it is exchanged, lent, borrowed, distributed, spent, and more. And, with money, there comes debt. With debt, there comes insolvency and, with it, bankruptcy and all the associated disasters.

Could we think of going a step beyond the institution of bankruptcy laws and imagine a financial system where people cannot go bankrupt? This is an idea that floats nowadays in the world's global consciousness. Perhaps the first proposal in this sense was made by Cory Doctorow in 2003 (during the pre-Facebook age) in his novel *Down and Out in the Magic Kingdom* [27], where he proposed a kind of "merit money," called "Whuffie," that people could accumulate on the good deeds that they performed. This money was a form of credit, but it could not be spent—it just produced perks and advantages for its owner. It was something that prefigured the "credit score" that Facebook and other social media would later develop. Maybe Doctorow was inspired by Mark Twain's story *The Million Pound Bank Note* (1903), where the protagonist finds that the mere possession of this banknote of enormous value entitles him with honors and goods without the need to spend it. But Doctorow may have been thinking of the concept of personal honor, fashionable in less monetized times than ours. As an honorable man you were entitled to privileges, but enjoying them did not mean that your honor would be reduced as a consequence.

Later on, the idea of using the credit score of social media as a form of money was proposed perhaps for the first time by Solitaire Townsend in 2013 [28]. The Chinese government seems to have taken the idea seriously with their plan of implementing a statewide system of social credit (*shèhuì xìnyòng iⓍxi*) [29] that would "grade" all Chinese citizens on a merit score. You get positive points for being a good citizen: helping an old lady crossing the street will bring you points from the lady and from the people who witnessed the deed. You get negative points when you do something bad, like getting a traffic ticket or just a bad report from someone who felt hurt by something you did. The Chinese social credit system can be seen as a form of money in the sense that it is based on the yin-yang opposition of debt and credit. For a Chinese citizen, having a sufficiently high social credit score is a prerequisite for being able to purchase certain things which, in the West, are possible only for the rich, plane tickets for instance. Something similar had been developed in earlier times in the Soviet Union, where the members of the Soviet Communist Party were considered as having a higher credit score than the others. They enjoyed non-monetary perks and services being par to of the *nomenklatura* system, not so different from what we call the "establishment" in the West.

A "reputation currency" could work, at least in a certain way. An advantage of such a system is that it may be rigged in such a way to create no negative credit (no debt). Could we eliminate the bad consequences of insolvency in this way? And, in a single sweep, we would eliminate such things as theft, robbery, corruption, swindles, and all the crimes related to money. Nobody ever could steal your credit rating at gunpoint! But, obviously, there are problems with the idea. Doctorow says about his creation, the "Whuffie" money, [30]

> Whuffie has all the problems of money, and then a bunch more that are unique to it. In *Down and Out in the Magic Kingdom*, we see how Whuffie – despite its claims to being "meritocratic" – ends up pooling up around sociopathic jerks who know how to flatter, cajole, or terrorize their way to the top. Once you have a lot of Whuffie – once a lot of people hold you to be reputable – other people bend over backwards to give you opportunities to do things that make you even more reputable, putting you in a position where you can speechify, lead, drive the golden spike, and generally take credit for everything that goes well, while blaming all the screw-ups on lesser mortals.

Reputation may be a terrible form of currency for those who find themselves at the wrong end of the scale. Have you ever been bullied as a teen? If you experienced that, you know how hard it can be to be the boy at the lowest rung of the ladder. It is known to be a cause of suicide among teenagers in Western Countries [31]. The only way to escape is to behave in the most abject way with the leaders of the group: flattering them and obeying their orders.

There exists at least one more case of a non-monetary currency system: scientific research. Scientists grade themselves on various scoring factors based on how popular their work is with other scientists, measured in various arcane ways, the most popular one being at present the so-called "h-factor". If you are a young scientist, your career depends on your credit score and that pushes you to conformism. You cannot afford to criticize your senior colleagues, nor to propose ideas or theories that are outside the commonly accepted wisdom in your field. That's a privilege you will earn only after getting your tenure and even then you will have to be careful about displeasing the powerful dons who control the funding of research.

The scientific ratings never go negative and, no matter how low the credit score of senior scientists can be, it is rare that they can be hit by the equivalent of a bankruptcy sentence. This is probably the reason why it is often said that "science progresses one funeral at a time." (a quote attributed to the German scientist Max Planck). It means that old scientists tend to block scientific progress until the natural phenomenon of biological collapse removes them from the system. It would be an interesting reform to introduce "negative points" in science and fire the scientists who publish one or more truly bad papers. But, before that happens, the "Whuffie trap" that Doctorow described would play its role to push scientists toward the most abject conformism. That would surely destroy that spark of creativity that, despite all odds, science has still managed to maintain up to now.

At this point, you can see that bankruptcy is not a bug but a feature of the system. It is one of the checks that the system has to maintain the link between the virtual entity that is money and the physical entities which are goods you can purchase. Like inflation, bankruptcy is an evolutionary tool that prevents the system from getting stuck in a no-win situation by removing the inefficient and obsolete entities which populate it. Were it not for bankruptcy, we would probably still have Blockbuster renting you video CDs and charging you if you are late in returning them. In the end, money may be a virtual entity and you may also define it as the devil's dung. But we are addicted to it and we keep playing the money game. Money is so deeply intertwined with the way our society works that we cannot even imagine how it could work without it. What could happen to us if a large financial collapse were to destroy the value of our mighty dollar? We cannot say for sure, but the mighty Globalized Empire might crumble like a house of cards in a single, huge, Seneca collapse.

Natural Disasters: Florence's Great Flood

Fig. 3.6 One of my books that survived the Florence flood of 1966. It is The Gold of Troy by Robert Payne. The illustration shows Sophia Schliemann, wife of the archaeologist Heinrich Schliemann, wearing the jewels found by her husband in the ruins of the city of Troy. You can see the dark spots of mud left by the water. The book still faintly smells of something undefinable but that was the typical smell of the time of the flood. (Photo by the author)

Not long ago, I was accompanying my daughter who was looking for an apartment in Florence. Since her family includes three cats, she needed a little garden and we were mainly visiting apartments on the ground floor. These places are always at risk of flooding and I was using a GPS app running on my cell phone to measure the height of the floor over the sea level. The employees of the real estate company accompanying us were often surprised and they would ask me what I was doing. At my explanations, they were bemused: flooding? In Florence? That can't happen! (Fig. 3.6).

These young men and women, typically in their thirties, had no personal memory of the great flood that hit Florence in 1966. They knew that it had happened, yes, but they were classing it as part of ancient history: barbarian invasions, the Black Death, the Crusades, and the like—events that took place in the remote past and that would not happen again. A flood of half a century before had no relevance in their daily planning.

It is the characteristic of natural disasters that they strike at intervals long enough for people to forget that they can and do strike. Flooding is one of those events and the 1966 flood of Florence is probably already beyond the forgetting line. But it was a major event: not the only case of a major flood affecting a modern city, but one that threatened to destroy the art treasures kept in Florence from the Renaissance. The flood affected many ancient buildings and damaged precious works of art, generating great concern all over the world. Fortunately, the number of casualties was relatively small.

I witnessed the 1966 flood as a 14-year old boy. It was one of those experiences that mark one's life even though my home was on relatively high ground and was not touched by the waters. But my father's office was downtown, at the ground floor, and it was invaded by the murky waters that filled it nearly all the way to the ceiling. Fortunately, there was nobody there when the flood arrived, but it was the place where I kept most of my books. Most of them were turned into heaps of mushy paper and I still perfectly remember the smell of gasoline or kerosene that these remnants of books emanated. Some could be restored and I still keep a few of them on my bookshelves.

The flood left a town in complete disarray. Nothing worked anymore: the shops had been flooded, the banks were closed, the sewage system was clogged with debris, there was no water in the buildings, no public transportation available, people's cars were soaked in mud and would not start, many homes were without electric power. And the Italian government, taken by surprise, was slow in bringing help.

In the days that followed, the Florentines rolled up their sleeves and started working. For those who experienced it, it was an incredible surge of community spirit and reciprocal help. The bad-smelling mud was shoveled away and there started the slow work of cleaning up and restarting. That also involved taking care of the flooded museums and ancient buildings with their art treasures. Soon, the cleaning effort ceased to be just a job for the citizens of Florence: people came from all over the world to help. They were called the "angels of the mud" and some of them were so taken in by the vibrant atmosphere of the reconstruction effort that they never left. They got married to Florentines and many of them are still there, getting old in Florence and taking care of their children and grandchildren, by now Florentines, too.

The story of the Florence flood has a happy ending: damage was limited and the city could be returned to its original conditions. It is not always like this: natural disasters are of many kinds and can cause much worse damage as well as horrific loss of human life. Floods, hurricanes, forest fires, earthquakes and other manifestations of Nature's force are rare events, but also common enough that each one of us is likely to experience one or more of them during our lives. Often, although not always, the distribution of natural disasters tends to follow the Pareto law, as we discussed in an earlier chapter. That is, they tend to behave according to a mathematical formula where the frequency of a disaster is proportional to its size raised to an exponent (power law). Disasters tend to be less probable the bigger they are but there is a non-zero probability that even extremely large events will occur.

In practice, on the basis of historical data, you may be able to say that a certain disaster has a specific probability to happen in your region, but that does not tell you when and where exactly it will take place. Imagine that the probability of, say, an earthquake of a certain size is 1% every year where you live. That means there is a 63% probability for the earthquake to strike within a century. But it might strike tomorrow morning, or after 99 years, or never over the next 100 years. It is nearly certain (more than 99.9% chances) that it will strike within the next 1000 years, but that helps you little in planning for this possibility. So, you have to plan taking into account the worst case hypothesis, which may be a good idea in general.

There exists is a whole taxonomy of rare natural phenomena that can do great damage to people. We may start with earthquakes. The kind that destroys buildings is rare but, when large earthquakes occur, the consequences are usually disastrous. The strongest ever recorded is "The Great Chilean Earthquake" which occurred on May 22, 1960, near Valdivia, in southern Chile. Its magnitude was measured as 9.5 on the Richter scale. It is an enormous value: the scale is logarithmic so that, in terms of energy, each whole number increase corresponds to an increase of about 31.6 times the amount of energy released. The Valdivia Earthquake did a lot of damage but, fortunately, few victims because it was preceded by a powerful foreshock that caused a lot of people to leave their homes before the main shock arrived.

There are many more examples of destructive earthquakes and everyone knows about the San Andreas fault that marks the two plates that form California and which slowly slide against each other in irregular bumps. The disastrous San Francisco earthquake of 1906 is a reminder of how dangerous living in California can be, but there were many more quakes in the area. Sometimes, it is said that California is waiting for "The Big One," an earthquake so powerful that everything West of the San Andreas fault would slide into the Pacific Ocean (or, alternatively, that everything East of the San Andreas fault would slide into the Atlantic Ocean). This is mostly folklore and media hype: it is true that earthquakes do occur in California and will keep occurring as the two continental plates keep moving, but there is no evidence that a humongous event, way larger than anything seen before, is brewing and will someday send San Francisco or Los Angeles to the status of park attractions to be visited by tourists wearing scuba gear.

California is just one sector of the great "Ring of Fire," a geologically active region that circles the Pacific Ocean. Japan, opposite to California across the Ocean, is part of the ring and another earthquake-prone, highly populated region. The great ring is just one of the many geologically active areas of the planet: those at risk also include the Mediterranean region, the Middle East, Central Asia and the Himalayan region, and more. An especially active region is the "Great Rift Valley" that goes from the Middle East to Central Africa. At the border between Somalia and Ethiopia, a geological process is in action to split the African Plate into two new separate plates. What's now a low-height valley will be a new sea, perhaps part of an ocean, but that will take millions of years.

There is no place on the whole planet that can be said to be completely free from earthquakes, but some places are surely quieter than others. In general, you are unlikely to experience major earthquakes if you live in the central areas of North America, in Australia, or in Eurasia, but note that a medium-sized earthquake struck Chicago in 1968. In general, the danger of earthquakes is nowhere so large that you should relocate far away from seismic areas, but surely you cannot ignore it if you live in one. In all cases, it is good practice to take precautions by living in a solid house—if you can—and to keep the emergency equipment that will be needed after the earthquake to cope with the disruption of services such as food, electricity, clean water, and more.

A phenomenon directly related to earthquakes is the *tsunami*, taking place whenever an earthquake shakes the seafloor. That can perturb a large mass of water that then moves across the ocean. When this water arrives near a coast, it takes the form of a wave, sometimes very large, that crashes on the shore and may destroy everything even for miles inland. The most tsunami-prone regions in the world are probably those on the coasts of the Pacific Ocean, along the great ring of fire, the most recent major tsunami in this region was associated with the Tohoku earthquake and it struck Japan in 2011. The Indian Ocean is also a tsunami-prone region and you may remember the 2004 tsunami that struck Indonesia, killing 230,000 people. The Mediterranean region is geologically active and it is also subjected to tsunamis: A relatively recent one struck the Italian cities of Messina and Reggio in 1908, causing a large number of victims. Much earlier, some 3600 years ago, a large volcanic eruption took place in the island of Thera (today called Santorini) in the southern Aegean Sea.

The related tsunami may have destroyed the Minoan civilization and generated the legend of the sinking of Atlantis.

The Atlantic Ocean is less active than other oceans in terms of moving tectonic plates, but it is nevertheless sensitive to tsunamis caused by coastal landslides. A source of a possible future Atlantic tsunami could be the collapse of a large section of the island of La Palma, one of the Canary Islands in the Eastern Atlantic. It could happen as the result of an eruption of the Cumbre Vieja volcano. According to some estimates, if this landslide were to occur, the result would be a wall of water up to 300 feet high moving across the Atlantic and reaching the East Coast of the United States in about nine hours [32]. The resulting damage on the coastal cities would be unimaginable, a true super Seneca Collapse. But we have no idea of when and whether such a disaster could take place.

If you live in a geologically active zone, you should also worry about volcanoes, probably the most destructive phenomenon generated by purely geological forces. A well-known example is the destruction of the cities of Pompeii and Herculaneum in Southern Italy in Roman times, in 79 CE. These cities were buried under a thick layer of ashes, excavations are still ongoing today and archaeologists keep finding traces of the bodies of the people who suffocated or were killed by heat, sometimes still in the position they had assumed when they died.

An even more spectacular case of a volcanic disaster is that of Toba, a "supervolcano" which erupted about 75,000 years ago at the site of present-day Lake Toba in Sumatra, Indonesia. It was perhaps the largest eruption known to have taken place in an age when our human ancestors had a chance to experience its effects. Some evidence indicates that the enormous mass of dust pushed into the atmosphere generated a "volcanic winter" that may have led to the disappearance of a large fraction of the human population of the time. That led to the "genetic bottleneck theory" proposed by a number of scientists [33] that would explain why the humans of today have a relatively small genetic differentiation. It may be because we are all descendants of the small group of people who survived the Toba eruption and who then spread all over the planet in an interesting case of "Seneca rebound." But, no matter how fascinating the bottleneck theory can be, at present it seems that the data do not support it [34]. Whatever the case, if a Toba-size eruption were to take place nowadays, it would likely destroy our whole civilization, maybe even causing the extinction of the human species. We can only hope that the "fat tail" of the Pareto distribution for volcanic eruptions is not so fat as to give a significant probability to such an event.

The Toba supervolcano eruption is related to plate tectonic activity [35] just like most of the volcanoes active today. But there is also another kind of volcano, related to the "hot spots" in the Earth's crust [36]. These volcanoes are generated by plumes of hot magma that start in the asthenosphere, the region of the Earth's mantle that lies just below the crust. These plumes look a little like the whirls and bubbles of a "lava lamp," although they move at an enormously slower speed. The flood of lava generated by a hot spot is often gentler and moves slower than the plate-generated kind—the "normal" volcanoes, because the magma is basaltic (or "mafic" using the geologists' jargon) and contains smaller amounts of dissolved gases than the "felsic" (again, the geologist's jargon) kind of volcanoes. So, the outflow of basaltic lava is less subjected to explosive outbursts.

A well known hot-spot volcano is the one that generated the Hawai'i archipelago over the past 5–6 million years, one island at a time, as the ocean floor moved over the hot spot (which may also have been moving). Today, you can see the hotspot in action at Kilauea, on the South—Eastern side of the Hawaii islands. The latest bursts of eruptions took place in 2018, it was not so gentle because it destroyed several homes and caused an earthquake, but no victims were reported. The Lōihi volcano, off the South-Eastern coast of Hawai'i, is the latest incarnation of the underground hotspot. It is presently undersea, but gradually growing and rising. It is expected to begin emerging above sea level about 10,000–100,000 years from now. It will surely be something spectacular to watch for our descendants who will have a chance to be there.

The Yellowstone hotspot volcano also deserves a mention [37]. Right now, it is quiet, it is not even a volcano. But, over the past 18 million years or so, the hotspot generated a succession of violent eruptions and floods of basaltic lava, at least a dozen of them were so massive that they are classified as "super-eruptions." The hotspot could become active again and generate a new supervolcano that could rival the ancient Toba in terms of global destruction, or even be much worse. It is another entry in the list of the event that could destroy the human civilization and even cause the extinction of the human species. But we cannot predict when (and whether) it will take place.

This review of giant natural disasters cannot neglect the possibility of meteorites, often called asteroids when they are large. Asteroids that fall on the Earth can cause enormous damage, so much that they are a popular subject of catastrophic movies. It is true, indeed, that the geological record shows several cases of large meteorites falling on the Earth's surface. An especially spectacular one was that of the Chicxulub meteorite hitting the Yucatan peninsula, Mexico, some 66 million years ago. The impact is commonly said to have caused the mass extinction that included the disappearance of the non-avian dinosaurs at the boundary between the Cretaceous and the Paleogene periods (K–Pg boundary). This idea had become almost universally accepted up to a decade ago, but it is now debated and often rejected: it seems clear that the dinosaurs were destroyed by a different phenomenon, a giant basaltic eruption that took place in the region we now call the Deccan, in India [38]. In any case, the risk associated with falling meteorites is extremely low and there are no reliable reports of anyone having been killed by one in modern times.

Geology-related disasters are often classed as "acts of God," meaning that they are completely unrelated to human actions, but this is not always the case. The human influence on the Earth system is by now so large that it affects even geological phenomena. For instance, the slow melting of glaciers caused by global warming, largely the result of human activities, is generating a phenomenon called "isostatic rebound" of the regions covered by ice caps. It works like this: the tectonic plate below the glacier "floats" over the fluid astenosphere, below the crust, just like all tectonic plates. The weight of the ice sheet on it pushes the plate down but, with the sheet thinning, the plate moves up. It is a very slow phenomenon but it destabilizes the whole area and may generate earthquakes, volcanoes, and sometimes tsunamis.

Many more natural phenomena are only partially natural: they may be triggered by human activity and the damage they generate may be increased by unwise human practices. Among these, we can list forest fires and hurricanes, often enhanced by global warming. A hotter atmosphere may make hurricanes more destructive, and it can also make forest fires more frequent and more deadly both because of the higher temperatures and because of droughts. In recent times, California has been struck by several major fires: these are natural phenomena but human activities can enhance their frequency and intensity in various ways. One is the change in the weather patterns caused by climate change, others are poor forest management practices. The "Oakland Firestorm" of 1991 is an example: the fire was enhanced by the introduction of non-native trees in California, the easily flammable eucalyptus trees [39].

Landslides are also triggered by human activities in terms of deforestation or poor soil management. A good example is the landslide that struck the town of Sarno, in Italy, in 1998, causing the death of 160 people, engulfed by a giant mass of slide coming down from the surrounding mountains. It was enhanced by the deforestation of the hills around the city. In some cases, landslides are wholly human-made: for instance, in 1966, the collapse of a pile of coal mining debris at Aberfan, in England, killed 116 children and 28 adults in a school that had been erected nearby.

Do we see any trends in the number and lethality of natural disasters? The data reported by *Our World in Data* [40] show that the sum of all reported disasters—of all kinds—reveals an increasing trend up to around the year 2000, then it starts going down. If we examine the data for different types of disasters, we see that phenomena as diverse as earthquakes, wildfires, and floods show this trend: their frequency goes up until the turn of the century and then declines or stabilizes. The trend for the number of fatalities is less clear-cut: some cases, such as for the deaths caused by extreme temperatures, we see an increase with the turn of the century, while for others, such as those caused by droughts, we see a clear decline starting from the 1920s and still ongoing. Finally, if you are worried about being struck by lightning, you may be happy to know that the data show that the number of fatalities in the US has been declining by a factor of almost 100 from 1900 to 2015.

These data are not easy to interpret: what made the frequency of many natural disasters go first up and then down? Did the Earth's weather patterns change? Or was it just a question of different reporting criteria? It is hard to say, mainly because the damage caused by natural disasters depends on several factors: it is not just the intensity of the forces of nature, but how people are prepared to cope with the event. So, we cannot know whether there will be larger changes in the coming decades: the ongoing global warming may make weather-related phenomena more destructive and more frequent, but that cannot be said with absolute certainty. What we can say is that, overall, natural disasters in the world have caused some 70,000 victims per year during the decade of the 2010s, so far. If the trend does not change, it means that, on the average, your probability to die struck by any of the several possible "acts of God," from floods to volcanic eruptions, is of the order of one in a million per year.

So, should you be worried? Yes, absolutely. First of all because, although the probability of dying is low, the probability of suffering heavy damage is way higher. Here, we may again remember the story of the statistician who drowned in a river of an average depth of 1.5 meters! If you were living in Florence in 1966, your probability to die because of the flood was about 0.003% (17 victims out of a population of ca. 500,000 persons), but almost everyone in Florence was negatively affected in various degrees. Then, your probability to die or suffer heavy damage in a major natural disaster depends very much on where you live. If you live in a mountainous area in a continental region, you should not be worried about tsunamis, unless you think of the movie *2012*, where the tsunami waves were taller than the Himalaya mountains! But if you live on an island of the Pacific or the Indian Oceans then, yes, a major tsunami has a significant possibility of striking you during your lifetime.

So, it makes sense to plan ahead for the possibility of major disasters. As usual with critical phenomena, it is not possible to predict where exactly a natural disaster will strike, nor how large it will be. That does not mean you should not apply the wise strategy proposed by captain Kirk of the Federation's starship *Enterprise*: "I never put myself in a no-win situation." It is a re-statement of the best strategy for winning at the Russian Roulette game: just don't play it! It is the strategy that I applied when my daughter was looking for an apartment in Florence for her family, using a GPS app to make sure that the apartment was high enough over the sea level that it did not risk being flooded in case of an event such as the 1966 flood. Maybe it will not happen during my daughter's lifetime, but why take chances? If you live in California, you should at least avoid buying a house that stands right across the St. Andreas fault or in the midst of a eucalyptus forest.

If you can't avoid living in dangerous areas, then your best bet for survival is to be ready. If you happen to face a forest fire racing toward your home, your hope consists in having your car ready and to have planned in advance the road to take to move away from the risk zone. It happened to a friend of mine who was living in Oakland, California, at the time of the firestorm of 1991. She was at home when she saw a giant wall of fire surging from the woods. She did not even have the time to put her shoes on, she ran for her life in her slippers, managing to start her car and outrun the firestorm. Then, if you live in the "Tornado Alley" in the central US, you should not just content yourself with the fact that the probability of being killed by a tornado is low even in that area, probably less than one in a million per year [41]. Most people living in that region equip their home with a "storm cellar," an underground refuge. You may never have to use it, but not having it is a risk not worth taking.

Overall, natural disasters are highly destructive but, mercifully, they are reasonably quick to go away, at least in their most intense form. After the earthquake has struck, the flood waters have retreated, the twister has faded in the clouds, there comes the moment to look around, assess the damage, and plan for rebuilding. Here, an important factor is scale. Small scale disasters, such as tornadoes and forest fires, are spectacular, but localized. In most cases, they may destroy a few homes, but the overall damage is limited. Then, if the people who have been hit have good insurance, they can rebuild their homes. This is what happened with the Oakland firestorm of 1991, in California: the fire burned to cinders some very expensive homes on the hills in the area, but when the time to rebuild them came there was no need of the kind of communal solidarity that the Florentines had shown in rebuilding their city, probably alien to the cultural orientation of the residents of the hills around Oakland [39]. Instead, most owners had insurance policies that allowed them to rebuild even larger homes, sometimes even extravagant ones. That was the case of my friend whose home had been destroyed in that fire: she and her husband were able to build themselves a better and bigger house. What they were most sorry about was having lost all the records of their previous life: the pictures of their marriage, of their children, of their families. Today, you probably have all those pictures in the cloud, so even in case of a wildfire destroying your home you will not lose the memories of a life. That is probably more likely to happen today if your cloud provider loses your records: it happened to the cloud provider MySpace that lost some 15 million of users' records in 2019 [42]. It is another kind of Seneca collapse, this one, fortunately, just virtual.

A different story is when the disaster is so large that the resources needed to rebuild are insufficient. An example of a disaster large enough to put the whole society to test is that of Hurricane María, in 2017. It was not an exceptionally strong hurricane and not even an unexpected one. It was just rain, rain, rain. Initially, it seemed that the effects had been limited, but the true size of the disaster became apparent months afterward. One reason was the poor response of the authorities but, really, the problem was that Puerto Rico was—and remains—*poor*. Not only poverty had weakened the island's infrastructure, but the poor lacked the extra resources which are needed when people have to recover from a catastrophe. On this subject, let me report an excerpt from Ariel Lugo's book *Social-Ecological-Technological Effects of Hurricane María on Puerto Rico* [6] (p. 49)

> Before María the consensus was to make government like a private enterprise, without realizing that the government tends to provide more benefits because it has a service mandate not a profit motive. Privatization makes money for entrepreneurs, lifts the economic status of the politicians that selected them, but often dramatically fails in public services to the citizens, particularly when faced with extreme events. A government operated according to the profit motive of the private sector will use cost-effectiveness as the criterion for action as opposed to public service and public good. The profit of the privatized agency or government sector is secured while portions of the public, which help to underwrite that profit, are left to fend for themselves.

Most people do not realize that, when examined objectively, government, not private entities, tends to deliver services most efficiently, that is, at less cost per unit benefit. And it gets a lot worse following an extreme event.

Here, Lugo hits a fundamental point: we are becoming more vulnerable to catastrophes with our emphasis on privatizing everything in the name of efficiency. In this way, we have no resources left to cope with extreme events nor to help people who are hit and cannot pay for what they need. We are making the social network tighter and more efficient, but at the expense of resilience. More than that, the very fabric of society is being destroyed because of the emphasis on efficiency: cooperation and trust among citizens disappear just when collaboration—rather than competition—becomes the fundamental virtue to attain the resilience we need to survive and rebound after the catastrophe. Lugo notes also that (p. 48),

> When faced with the overwhelming effects of an extreme event, the human spirit and will rise to the occasion. Many Puertoricans did not wait for external aid and choose instead to rise to help themselves and their neighbors.

Could this kind of resilience be planned for even before a disaster will strike? Some people appear to be engaged in this kind of planning. In the US, there is the movement of the "preppers," or "survivalists," people preparing for whatever major disaster may strike them, including the end of the world as we know it (TEOTWAWKI). In many cases, preppers emphasize individual or single-family preparation rather than community resilience, they may stockpile food, supplies, and weapons in their cellars in the expectation for the worse to come. A different approach may be common in Europe with the "Transition Towns" movement [43] which emphasizes collective action to preserve the local social network. These are experiments in building community-level resilience by means of collaboration, local resources, local agriculture, and sometimes using local currencies. It does not seem that survivalists or transition towns people have been put to test yet by a true emergency situation, so we do not know how well these ideas will withstand contact with reality. In principle, both ideas may be good in some circumstances but, as usual, we move into the future without being sure that we are taking the right direction.

Mineral Collapses: The Coming Oil Crisis?

Fig. 3.7 The author, Ugo Bardi, discussing the future of the oil industry at a meeting of the Club of Rome in Vienna, 2017. You can see on the whiteboard that his prediction was not very optimistic: it is the Seneca curve

In 2003, I attended my first conference on oil depletion in Paris. There, I met the larger-than-life figures of the experts who had revamped global interest in oil depletion and founded the Association for the study of peak oil (ASPO): Colin Campbell, Jean Lahérrere, Ali Morteza Samsam Bakhtiari, Matthew Simmons, and many others. In Paris, everything looked new, remarkable, exciting: we were riding a wave of interest in oil depletion that had started in 1998 with an article by Campbell and Laherrere in *Scientific American* [44]" titled "The End of Cheap Oil." The resonance of that article had been enormous: among harsh criticism and enthusiastic acceptance, the term that Colin Campbell had coined, "peak oil" had rapidly gained worldwide popularity.

For me, the Paris conference was the start of my interest in collapse. True, the peak oil concept, did not imply that decline was to be faster than growth. But, already in 2005, I published my first paper on oil depletion [45] finding the conditions that led to what I called "sawtooth-shaped" collapsing curve. The idea of calling it the "Seneca curve" came much later. I was not the only one who found the concept of peak oil fascinating. The importance of oil as the main support of civilization was well known, but the idea that oil was becoming scarce provided a new interpretation of past events, from the great oil crisis of the 1970s to the 2001 attacks against the World Trade Center in

New York. Peak oil had a certain ring of apocalypse to it, especially because many people understood the peak as the same thing as running out of oil. Not everybody misunderstood the concept so badly, but peak oil, it was said, meant the end of the world as we know it and we had better be aware of the punishment that the dark divinities of the black liquid found underground were preparing for us.

The popularity of the concept of peak oil rose to high levels in the early 2000s, but it was short-lived. It may have peaked around 2006 then, way before it could be said that any of the peak oil forecasts had been right or wrong, it started declining [46]. Not even the great oil price spike of 2008 generated more than a transient blip of interest. In time, a new wave of optimism came and the concept of peak oil became politically unnameable, sometimes a source of scorn for those who still dared to propose it.

The peak oil parable is just an example of how human worries tend to go in cycles. Erwin Schlesinger, former US secretary of state, said that people have only two modes of operation: complacency and panic. It may also be that these two modes tend to go in cycles, periodically replacing each other. So, the wave of interest in oil depletion that had started in 1998 was not the first: the idea had ebbed and flowed all along the great cycle of exploitation of crude oil. Already in the 1950s, the American geologist Marion King Hubbert had proposed his "bell-shaped curve", generating an early cycle of interest that faded in the 1980s with the wave of enthusiasm for the Internet and the dot-com economy. It may very well be that the current complacency phase could give rise to a new phase of panic in the near future. And, in the case of crude oil, the term "panic" is justified. Without liquid fuels, everything would stop in the world. Recently, Alice Friedemann published a study on this subject: *When the Trucks Stop Running*, [47] and the title, alone, tells the whole story. No fuels, no trucks, no food, no civilization. Could it really happen?

It could. Something similar already happened with the great "oil crisis" of the 1970s that for a period seemed to destroy the very foundations of the Western civilization. If you experienced that crisis, you cannot forget what happened: gas prices suddenly skyrocketing, long lines at the gas stations, governments enacting all sorts of measures: lower speed limits on highways, "odd-even rationing" schemes, support to the production of small cars, and more. The shock on the financial system was even worse: recession and two-digit inflation. It was a disaster for a world that had experienced, up to then, more than 2 decades of uninterrupted economic growth. The data show how world oil production had started declining faster than it had been growing before the peak. It was a clear case of a Seneca curve (Fig. 3.8).

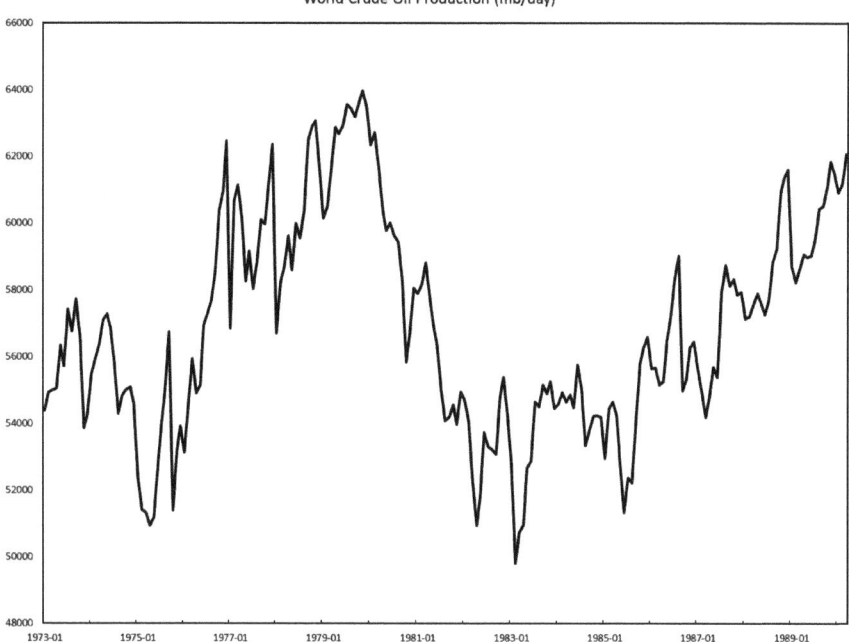

Fig. 3.8 Oil production at the time of the great oil crisis of the 1970s. Data from IEA

Eventually, the crisis that had started in the 1970s abated. With the development of new oil fields such as those of the North Sea, production started to grow again. With the mid 1980s, pre-crisis production levels were reached again and then surpassed. In the decades that followed, the world oil market turned out to be remarkably resilient: we saw wars, collapses, international crises, and all sorts of changes and disasters. But crude oil and natural gas kept flowing everywhere in the world.

Today, the events of the 1970s are part of the "memory fog" of humankind, a fog that turns into ancient history everything older than a few years (or even less than that). So, the story of the oil crisis was turned into something that looks like an ancient myth of good vs. evil. The way it is often told, it involves a group of power-hungry Arab sheikhs (or maybe ayatollahs) who had attempted to take over the world using oil as a weapon. But their efforts were eventually thwarted by the good people of the West who found new sources of oil. From then on, everything had been well in the best of worlds.

There are some elements of truth in this simplified version of the story. If we look at the global production data over the past century or so, we can see how the increase has been nearly continuous. True, the chart is optimistic because it reports volumes produced and not energy—which is what we are interested in. But, overall, the growth of oil production is a real phenomenon (Fig. 3.9).

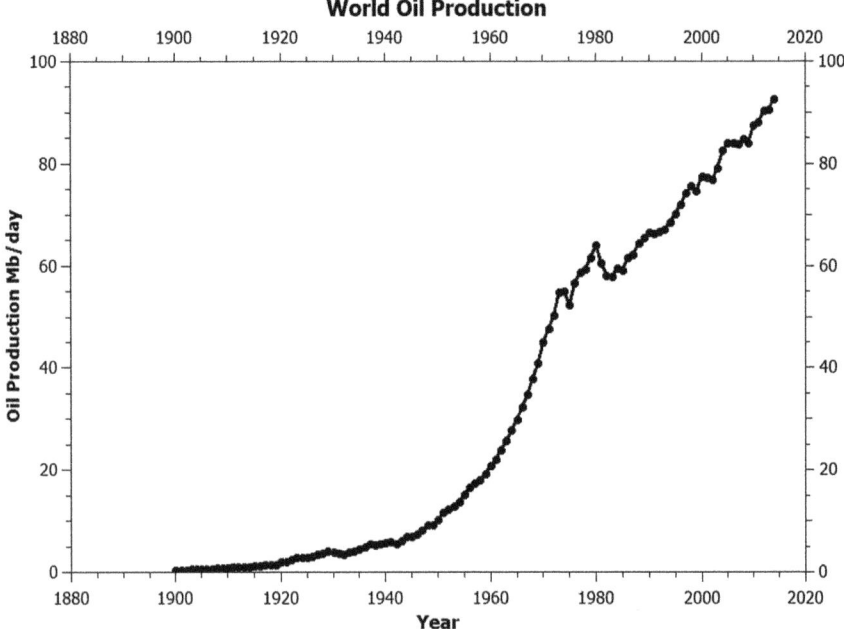

Fig. 3.9 World oil (all liquids) production. Data from "The Shift Project" https://theshiftproject.org

But there remains in our collective consciousness a deep unease that derives from the realization of how fragile our prosperity is. Not for nothing was the so-called "Carter Doctrine" expressed during the oil crisis years. It stated that the interests of the Middle East regions are vital to the United States and that the US will consider all attacks to these regions as a threat to its national security. There is a logic in this attitude: a large fraction of the oil reserves of the world is located in this region. If something goes wrong with the oil production of one of the major oil producers of the Middle East, Iraq, Iran, and Saudi Arabia, it will affect not just the United States but the whole world. It seems that the world's energy security hangs on political factors that may suddenly create unexpected problems: this was what happened with the great oil crisis of the 1970s. And the question is: could it happen again?

To answer this question, we can start from a favorite sentence by Colin Campbell, one of the first proponents of the "peak oil" concept, "the availability of crude oil today depends on events that took place during the Jurassic period and that cannot be influenced by politics." In other words, the supply of oil is finite despite some politicians claiming the opposite [48] and also despite the efforts of a group of vocal contrarians who try to push the concept that oil resources are really infinite, being continuously recreated by mysterious "abiotic processes" operating in the depths of the Earth [49]. It does not work that way, if you are an adult you should know that after you have eaten your cake, you don't have it anymore.

The oil industry seems to be perfectly aware of the limitations of available resources and spends considerable efforts on estimating the size of the available oil "cake". Obviously, these efforts are stimulated by the fact that resources are a factor in attracting investments. As you can imagine, it is not an easy task to evaluate the amount of something that lies miles underground. But there exist sophisticated measurement technologies that, coupled with even more sophisticated statistical treatment of the data, allow the industry to perform reasonably accurate estimates of the resources that are expected to lie hidden underground.

The problems with resource estimates is not so much a technical one but a political one: the search for a top position in the pecking order in the oil world may lead some governments or company boards to "adjust" the results of their analysis. In a 1998 paper, Colin Campbell and Jean Lahèrrere noted how the estimates for the oil reserves of six Middle Eastern countries members of OPEC showed an abrupt bump upward in the mid-1980s, creating a total of 300 billion additional barrels of oil added without having reported major discoveries of new fields. One can at least suspect that the estimates had been tweaked, and not a little, for political purposes. Western governments are not immune from exaggerated claims. As an example, much resonance was given in the media in 2016 about the "discovery of new oil reserves" in Texas, about 20 billion barrels calculated as worth some $900 billion [50]. One problem was that it was not a "discovery" but simply a new estimate of the quantity of technically recoverable oil in known deposits. But there was a much bigger problem that Arthur Berman noted [51]:

> Where did the $900 billion value come from? Multiply 20 billion barrels times $45 per barrel and you get $900 billion. In other words, if the oil magically leaped out of the ground without the cost of drilling and completing wells; if there were no operating costs to produce it; if there were no taxes and no royalties.

> According to the USGS' input data, it would take 196,253 wells to produce the 20 billion barrels if it exists. At $7 million per well, that would cost almost $1.4 trillion in drilling and completion costs alone.

> It would cost more than $1.4 trillion to generate $900 billion in revenue resulting in a net loss of $500 billion at $45 oil prices excluding all operating expenses, taxes and royalties–and no discounting.

That's a discovery that no one can afford to make.

But there is an even bigger problem with reserve estimates. Assuming that they are correct, they tell you something about the volumes that you may be able to extract, but nothing about the cost of extraction. As you may imagine, that is more than a small problem: it is like evaluating the military power of a country simply by counting the number of soldiers it can field, neglecting their firepower and their willingness to fight. That is a mistake Saddam Hussein made when he tried to hold onto Kuwait in 1991, just an example among many. The fact that some poor guys with rifles stand in a trench, somewhere, does not mean that they will be able to fight effectively and, in the same way, the fact that some "extractable oil" exists, somewhere, does not mean it will be extracted, unless somebody will be willing to pay the costs involved.

Despite the technical sophistication they deploy in the task, oil companies —just as all mining companies—seem to have little or no interest in using models that take into account the costs of extraction of mineral reserves to estimate future production. The most sophisticated model they normally use to peek into an uncertain future is the "reserve/production ratio" (R/P). It works by dividing the current estimated amount of reserves by the yearly production rate. The result is a number that can be interpreted as the number of years that production could go on at the current rates before the resource runs out completely.

The reason why companies (and politicians, too) love the R/P ratio is that it normally provides a comfortably large number of years before we run out of anything. For oil, for instance, the R/P ratio stands today at some 50 years, that for coal at a few centuries or, in some assumptions, at more than a thousand years. Most people understand from these data that there is nothing to be worried about regarding oil for at least 50 years and, by then, it will be someone else's problem. And, if we really have a thousand years of coal, then what is the fuss about? Add to this the fact that the R/P ratio has been increasing over the years and you understand the reasons for a rather well-known statement by Peter Odell, who in 2001 said that we are "running into oil" rather than "running out" of it [52]. In this vision, extracting a mineral resource is a little like eating a cake. As long as you have some cake left, there is nothing to be worried about. Actually, the peculiar cake that is crude oil has the characteristic that it becomes bigger as you eat it.

But then, if there is still plenty of cake to eat today, surely there was even more of it at the time of the great oil crisis of the 1970s. Then, maybe it is true that the crisis was all the fault of those evil Arab sheikhs, wasn't it? Again, adult people should recognize that blaming one's problem on some evil characters is not the best way to solve it.

As is often the case with complex systems, the oil crisis of the 1970s was a complex phenomenon generated by a chain of feedback effects. Depletion was a trigger that started the chain, but it was not in itself the cause of the disaster. And the same is true for political factors. They were not the "cause" of the disaster, not any more than a feather was the cause of the broken camel's back, as in the proverb. The oil crisis of the 1970s was a problem of the size of the faucet, not of the tank. No matter how much oil there was, somewhere underground, the capability of the industry to extract it was insufficient to satisfy the a growing demand. It is a problem that Arthur Berman perfectly framed when he said that considering only the underground oil resources is as "*if the oil magically leaped out of the ground without the cost of drilling and completing wells; if there were no operating costs to produce it; if there were no taxes and no royalties*" [51].

More precisely, the problem in the 1970s was that the industry was unable to keep enlarging the faucet at an exponentially increasing rate, as had been the rule from ca. 1940 to 1970. During that period, production was doubling every 10 years and, indeed, in some 30 years it had increased by a factor of nearly 10. If it had continued doubling worldwide at the same rate up to now, it would have doubled 5 times more and today the oil industry would produce about 30 times more oil than it did in 1970. Starting from about 50 million barrels per day, production would have arrived today at the fantastic value of one billion barrels and a half per day—while in the real world it is less than 100 million barrels.

Of course, that could not happen and it did not happen. Not only was it physically impossible to keep production growing for such a long time, but we would all have been cooked well done by global warming in the meantime. So, the oil crisis was not a bug but a feature: it was a needed adjustment in order to slow down the system and make it compatible with the real world. This interpretation is confirmed by the fact that it had been predicted in advance on the basis of exactly these considerations. In the 1970s, Pierre Wack and his group at Shell Oil were applying a technology called "scenario analysis" to oil production and they had noted that the evolution of the oil market was leading to a completely different situation from the one that had been standard during the past decades. He wrote in 1985 [53] that

- The oil market—long characterized by oversupply—was due to switch to a sellers' market.
- Soon there would be virtually no spare crude oil supply capacity.
- Inevitably, the Middle East and, in particular, the Arabian Gulf would be the balancing source of oil supply.
- The great demand on Middle East production would bring a sharp reduction in the Middle East reserve-production ratio, if met.
- The sharp peak in Middle East production would not be allowed to occur. Intervening factors would include a desire by Arab countries to extend the lifetime of their one valuable resource and a cornering of the world energy market by Gulf producers for perhaps 10 to 15 years by limiting production.
- Only something approaching a sustained worldwide depression could reduce the growth of demand for Middle East oil to levels where the anticipated sellers' market would be too weak to command substantially higher oil prices.

Wack wrote about what the consequences would be:

More than 20 centuries ago, Cicero noted, "It was ordained at the beginning of the world that certain signs should prefigure certain events." As we prepared the 1973 scenarios, all economic signs pointed to a major disruption in oil supply. New analyses foretold a tight supply-demand relationship in the coming years.

Now we saw the discontinuity as predetermined. No matter what happened in particular, prices would rise rapidly in the 1970s, and oil production would be constrained—not because of a real shortage of oil but for political reasons, with producers taking advantage of the very tight supply-demand relationship. Our next step was to make the disruption into our surprise-free scenario. We did not know how soon it would occur, how high the price increase would be, and how the various players would react. But we knew it would happen.

In the 1960s, oil reserves were considered abundant and no supply problem was foreseen for the short and medium term future. The problem was how to find the financial resources needed to keep production growing as it had been growing during the previous decade. Today, the situation is similar but worse: the problem is how to find the financial resources needed to keep production at least as stable (in energy terms) as it has been during the past decade. In both cases, the task was not and is not impossible, but it is surely difficult. In 1973, the relatively minor geopolitical shock of the Arab-Israeli war sent the system tumbling down the Seneca Slope. Today, another geopolitical shock could have the same effect.

There are factors, today, that could create a new oil crisis, possibly much worse than the one that started in the 1970s. On the demand side of the market, the fossil fuel industry is threatened by several factors. Renewables such as solar and wind, already produce energy at lower costs than those of fossils and that may be pushing coal toward extinction. Changes in the transportation market are also changing the rules of the game. Liquid fuels are mainly used in transportation: typically a good 50% of the oil industry production is gasoline. To this, you may add about 20% of diesel fuel and the result is that some 70% of the output of the industry is for internal combustion engines used for transportation. This market is threatened by two factors: one is the diffusion of electric vehicles, the other the diffusion of the concept of "Transportation as a Service" (TAAS) [54]. Electric cars can be powered using electricity produced by any source and the preference will reasonably go to renewable energy since it is clean and inexpensive. TAAS, then, may make individual cars as obsolete as wearing coats made of home-tanned bear skins. The concept of TAAS is not necessarily based on electric vehicles, but it may surely reduce the number of cars on the road, promote more efficient vehicles and a more efficient way of using them. The final result is likely to be a reduction in the demand for oil products.

These factors may badly dent the market of the oil industry as the result of the "collapse of the demand," a term that seems to be more acceptable in the mainstream debate than the tainted "depletion" term. But depletion is also gnawing at the profits of the fossil industry. No matter how enthusiastic one may be about the shale oil "miracle," it is a fact that miracles do not exist and that depletion is going to make the extraction of crude oil, gas, and coal more and more expensive as time goes by. Several producing regions have already gone through their Hubbert peak, often plunging into wars and social unrest as the result. Many of the main current wars are located in regions that saw the decline of national oil production: Yemen, Syria, and Venezuela are just some examples.

Eventually, it does not matter so much whether the problem is related to supply or to demand, these are two sides of the same coin. Nothing is produced at a price too high for customers to pay for it, and customers will never buy something they cannot afford. So, the destiny of the oil industry may well be to be brought down in a spectacular collapse by the candle burning at both ends: depletion on one side, demand decline on the other—it is another typical example of the concept of "dynamic crunch" that generates the Seneca cliff. We do not need a large reduction in the demand for transportation fuels to generate a spiral of decline for the oil industry. Less demand means less production, less production means the loss of economies of scale, and the loss of the economies of scale means higher costs that translate into higher prices which also depress the demand. And so it goes until it hits the bottom. As Lucius Annaeus Seneca said, long ago, "ruin is rapid."

So, there is a significant probability of seeing an oil shock scenario playing out in the near future (Fig. 3.7). It may be triggered by a decline of shale oil production in the US, maybe coupled to a political shock reducing the export capabilities of other producers, such as Saudi Arabia or Iran. The results would be similar to those seen in the 1970s: oil prices would skyrocket, the economies of industrial countries would go into recession, importing states would need to implement measures for reducing oil consumption. Although today we are not so dependent on oil for electrical energy production as we were in the 1970s, we are still highly dependent on liquid fuels for transportation—with about 85% of oil production being used for fuels. Actually, we may well be *more* dependent on oil for transportation today because, everywhere, the tendency toward urban sprawl has generated suburban agglomerates of homes and shopping centers which can hardly be serviced by public transportation. So, a new oil shock would generate again long lines at service stations and fuel rationing might become necessary. The new oil shock might well be much more destructive than the earlier ones, also considering that today we lack the equivalent of the brand new oil fields of the North Sea coming to the rescue, as was the case in the 1970s.

But a new oil crisis may not be a bad thing, either. Since we have been consistently unable to curb the consumption of fossil fuel in order to reduce carbon emission, it may be that peak oil or, more exactly, the Seneca Collapse of the oil industry, would solve the climate problem by having emissions crashing down before the ill-fated "2 degrees" threshold is reached and surpassed. And if that is not enough? That is, what happens if the threshold has already been passed and we are facing the dreaded climate tipping point leading toward a runaway climate change? It that case, both the current sects of catastrophists will be satisfied: we would die in fire *and* in ice, and for sure that would suffice!

The Seneca Cliff and Human Violence: Fatal Quarrels

Fig. 3.10 Tokyo after the bombardment of 9/10 March 1945. Photo taken by Ishikawa Kōyō (1904–1989) https://en.wikipedia.org/wiki/Bombing_of_Tokyo_(10_March_1945) #/media/File:Tokyo_kushu_1945-4.jpg

Years ago, I was somewhere in central Tokyo, in a place where I could have a good view of a large expanse of roads, squares, and areas where new skyscrapers were being built, a rare sight in a city that doesn't seem to value its own skyline as a sightseeing attraction. Sitting on a bench, nearby, there was an old Japanese man. Maybe because I was an obvious *Gaijin*, a foreigner, he endeavored to tell me in a mix of Japanese and English what the place around us was when he was a young man, just after the end of the Second World War. At some moment, he made an arching gesture with his arm, as to encompass the whole city, and said something like, "all destroyed, nothing, nothing, all the same, *mina onaji…*" I knew what he was referring to: I had seen pictures of Tokyo after the fire-bombing in 1945 and it was exactly what this old man was describing. The allies had used incendiary bombs against the mainly wooden houses of the town: the fires had not only flattened everything, but had left no chance to the inhabitants who found themselves trapped without ways to escape. In Tokyo, firebombs killed some 100,000 civilians in a single bombing raid on the night of 9/10 March 1945 (Fig. 3.10).

Over the years, a brand new Tokyo has been built over the ruins of the destroyed one, but everywhere the city still gives you a certain sensation of impermanence, something like living in the world of Basho's poems. Every time a small earthquake shook the building of the University of Tokyo where I was working at that time it was a little like hearing good old Godzilla stomping its giant feet just around the corner. And it is difficult to walk in Tokyo and miss that the large avenues crisscrossing the city blocks have a purpose. They were designed to act as barriers against fires spreading in case of a new wave of incendiary bombing.

The cities of the Western world have been free from aerial bombardments for more than half a century by now, with only some exceptions, such as Belgrade in 1999. But some people in Europe are old enough to remember the daily raids, the rushes to the bomb shelters, the flashes, the smoke, and

Fig. 3.11 The author in front of the ww2 air raid shelter still standing in the garden of his home, in Florence. It is a reminder of times past, but it cannot be excluded that it could become useful again in the future

the terrible noise of the dropping bombs. When they will be gone, no living memory will remain of those moments, but some physical memory will, for instance in the form of faded "bomb shelter" signs on some old buildings. The garden of the house where I live now still keeps an ogival concrete shelter used during WW2 by the inhabitants to defend themselves against the fires and the splinters. It is so heavy that nobody knows how to get rid of it and so it is still there. Maybe it will turn out to be useful again in the future, who can say? refers to Fig. 3.11.

Today, we are trying hard to forget what war is, but it remains with us, a ghost that we seem unable to exorcise. There is a quote attributed to Leon Trotsky that goes, "You may not be interested in war, but war is interested in you." Trotsky probably never said that, but it is a good description of the fact that when a war starts you have little or no possibility to avoid being affected by it.

Lev Tolstoy was among the first to speculate about the reasons for war when he wrote in his novel War and Peace (1867):

> To us it is incomprehensible that millions of Christian men killed and tortured each other either because Napoleon was ambitious or Alexander was firm, or because England's policy was astute or the Duke of Oldenburg wronged. We cannot grasp what connection such circumstances have with the actual fact of slaughter and violence: why because the Duke was wronged, thousands of men from the other side of Europe killed and ruined the people of Smolénsk and Moscow and were killed by them.

> To us, their descendants, who are not historians and are not carried away by the process of research and can therefore regard the event with unclouded common sense, an incalculable number of causes present themselves. The deeper we delve in search of these causes the more of them we find; and each separate cause or whole series of causes appears to us equally valid in itself and equally false by its insignificance compared to the magnitude of the events, and by its impotence—apart from the cooperation of all the other coincident causes —to occasion the event. To us, the wish or objection of this or that French corporal to serve a second term appears as much a cause as Napoleon's refusal to withdraw his troops beyond the Vistula and to restore the duchy of Oldenburg; for had he not wished to serve, and had a second, a third, and a thousandth corporal and private also refused, there would have been so many less men in Napoleon's army and the war could not have occurred.

> <.. > Without each of these causes nothing could have happened. So all these causes—myriads of causes—coincided to bring it about. And so there was no one cause for that occurrence, but it had to occur because it had to. Millions of men, renouncing their human feelings and reason, had to go from west to east

to slay their fellows, just as some centuries previously hordes of men had come from the east to the west, slaying their fellows.

<.. > it was necessary that millions of men in whose hands lay the real power—the soldiers who fired, or transported provisions and guns—should consent to carry out the will of these weak individuals, and should have been induced to do so by an infinite number of diverse and complex causes.

<.. > When an apple has ripened and falls, why does it fall? Because of its attraction to the earth, because its stalk withers, because it is dried by the sun, because it grows heavier, because the wind shakes it, or because the boy standing below wants to eat it?

<.. > And so there was no single cause for war, but it happened simply because it had to happen

Tolstoy was no scientist, but these words could have been written by a modern scientist versed in system science. It is a characteristic of complex systems that their behavior can hardly be described in terms of "causes" and "effects," rather, they change, move and evolve as the result of the interplay of forcings and feedbacks. This was the intuition of Tolstoy, who had not seen the 1812 Patriotic War in person, but had been with the Russian army during the Crimean War (1853–1856) and the siege of Sevastopol in 1855. That war is today mostly forgotten but it provides another example, if needed, of a totally useless conflict. Why was it fought? Apart from some silly pretexts about the freedom of cult of some religious sects, nobody seemed to know for sure at that time and you would hardly find anyone, today, who could explain it either. Nevertheless, the Crimean war prefigured the much larger and more cataclysmic wars of the 20th century, so much that we could rightly call it "World War Zero" [55].

Many studies and assessments of war as a social phenomenon have been published after Tolstoy and, today, we seem to regard war with a certain degree of optimism, perhaps because the world has not seen another large world war after WWII for some 60 years, so far. Among the most optimistic assessment, we have the one by Steven Pinker with his well-known book, *The Better Angels of our Nature* (2011) [56]. Pinker's thesis is that modern times became less violent during the past decades, and that it is a trend that will be maintained in the future. Other historical analyses of war are also optimistic.

According to Rudolph Rummel (1932–2014), democracies are much less likely than dictatorships to engage in wars [57]. In this interpretation, promoting democracy could be a good way to avoid wars and the trends toward more democracy in the world could be a reason why we may be living in less troubled times than in the past. It may be because of Rummel that the idea of "exporting democracy" has become so popular, nowadays, although in ways that leave many of us a little perplexed.

In any case, both Pinker and Rummel base their conclusions on historical data and may well be right for the time range they consider: the past few decades or, at most, the 20th century. It is true that the "big one," the Second World War, was probably the most destructive war in history and that afterward there were no more wars of comparable size. But is that a true long-term trend or just a statistical fluctuation? [58]. The current world's political situation does not seem to provide ground for optimism, with reciprocal threats of nuclear annihilation being again exchanged nowadays, as it was fashionable to do in the 1950s.

To understand what we are facing, we need data that go beyond the past few decades and, as much as possible, beyond the past century. The task of analyzing wars from a long-term statistical viewpoint was first attempted by the British physicists Lewis Fry Richardson (1881–1953). Richardson was in many ways well ahead of his age, and his contributions in fields such as meteorology and fractal analysis were so advanced that it took time for them to become part of mainstream knowledge. He was also a pacifist who tried to understand what generates wars and how we could, perhaps, avoid them. So, he performed a series of analyses of the frequency and the size of human wars and more in general of what he called "fatal quarrels," those human interactions ending with the death of someone.

Richardson proposed that wars and homicides tend to follow a "Poisson distribution," [59]. In time, it was found that wars are another kind of critical phenomenon [60–62]. Just as earthquakes and wildfires, wars tend to follow power laws. The initial intuitions of Richardson were confirmed by later studies. Let me show you some data from the database prepared by Brecke [63], covering some 600 years of human history. Together with my coworkers, Martelloni and Di Patti, we analyzed these data in a recent paper [62] (Fig. 3.12).

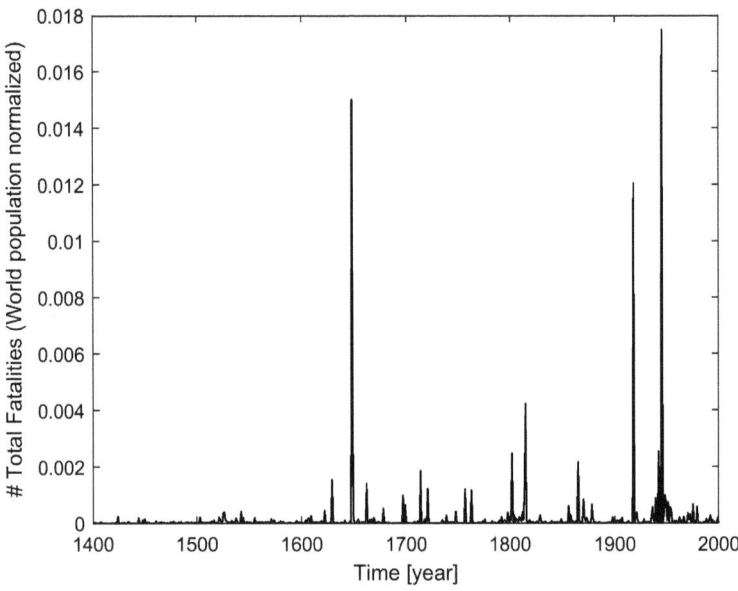

Fig. 3.12 Total War Fatalities in the world, normalized to the world population. Data from [62]

You see how the history of wars is dominated by a few very large conflicts —shown as normalized for the world population in the figure. The scene seemed to be relatively quiet up to mid-17th century, then a series of war "spikes" started. Some are especially large and recognizable: the 30 years war, the Napoleonic campaigns, the Crimean war, the First and the Second World War. It is hard to see, here, any continuous trend: what we can say is that, over 600 years of wars, the absolute number of wars has increased, but it decreased if we normalize it to the increasing world's population. Then, if we look at the frequency of wars as a function of their size, we find the typical "power law" distribution of critical phenomena. That is, large wars are less frequent than small ones, but there exists a "fat tail" in the distribution that makes large events not as unlikely as they would be if they were purely casual —or random—events.

As argued, among others, by Clauset [61], the so-called "long peace" of the period after the Second World War is not statistically significant as a change in past trends. Clauset arrived at the conclusion that a war the same size as the Second World War has a more than 40% probability to occur within the next 100 years, while a war with one billion battle deaths and, presumably, the extermination of most of humankind, has a median waiting time of little more than 1000 years. That is, it is nearly certain considering that human beings have been waging war against each other for longer than that.

There follows that wars, it seems, are emergent phenomena in the complex social system formed by human groups. In other words, it is not the will of mad rulers that generates wars but some kind of collective force that emerges out of a social network as the result of reinforcing feedbacks. War appears to be an unavoidable consequence of the behavior of human beings, perhaps a result of our primate ancestry [62, 64]. It is remarkable how this quantitative analysis validates the intuitions that Lev Tolstoy proposed one century and a half ago.

Of course, I said more than once in this book that predicting the future by extrapolating from past trends is dangerous and unreliable. Yet, the results we found over 600 hundred years of history are sobering: if nothing changes in the behavior of humankind or in the structure of society, the probability of major wars occurring in the near future is high. And we are not extrapolating anything if we just look at the current trends: wars are going on right now and the behavior of the "great powers" seems to be increasingly aggressive and reckless in a situation that reminds one more and more of what preceded the First World War. At that time, it is likely that nobody among the leaders had any idea of what the consequences of their decisions were. It was said that WWI was to be the war that would end all wars but, judged in retrospect, that looked a little optimistic. And yet, the same concept, the war that was to end all wars, was repeated as recently as with the 2003 invasion of Iraq.

Is there some way to stop wars in the future? There is no lack of ideas on the matter and it may be interesting to quote here the book by David Wilkinson, *Deadly Quarrels* (1980).

> The most common way of contributing to the debate over war causation and peace strategy has been to assert some definite theory, to show how it fits current circumstances, and to deduce immediate practical conclusions. If we follow this public debate, we may expect to be told that war is a consequence, for instance, of wickedness, lawlessness, alienation, aggressive regimes, imperialism, poverty, militarism, anarchy, or weakness. Seldom will any evidence be offered. Instead, the writer is likely to present a peace strategy that matches his theory of war causation. We shall therefore learn that we can have:
>
> - Peace through morality. Peace (local and global) can be brought about by a moral appeal, through world public opinion, to leaders and peoples not to condone or practice violence, aggression, or war, but to shun and to denounce them.
> - Peace through law. Peace can be made by signing international treaties and creating international laws that will regulate conduct and by resorting to international courts to solve disputes.

- Peace through negotiation. Peace can be maintained by frank discussion of differences, by open diplomacy, by international conferences and assemblies that will air grievances and, through candor and goodwill, arrive at a harmonious consensus.
- Peace through political reform. Peace can be established by setting up regimes of a nonaggressive type throughout the world: republics rather than monarchies; democratic rather than oligarchic republics; constitutionally limited rather than arbitrary, autocratic regimes.
- Peace through national liberation. Peace can be instituted only through the worldwide triumph of nationalism. Multinational empires must be dissolved into nation-states; every nation must have its own sovereign, independent government and all its own national territory, but no more.
- Peace through prosperity. Peace requires the worldwide triumph of an economic order that will produce universal prosperity and thereby remove the incentive to fight. Some consider this order to be one of universal capitalism, or at least of worldwide free trade; others hold it to be some species of socialism, reformist or revolutionary, elitist or democratic.
- Peace through disarmament. Peace can be established by reducing and eventually eliminating weapons, bases, and armies, by removing the means to make war.
- Peace through international organization. Peace can be established by creating a world political organization, perhaps even a constitutional world government resembling national governments, to enforce order and promote progress throughout the world.
- Peace through power. Peace can be maintained by the peaceable accumulation of forces, perhaps overwhelming, perhaps preponderant or balancing or adequate-sufficient to deter, defeat, or punish aggression.

It is clear that we are not going anywhere if we are dealing with nine different and incompatible theories on how to establish peace. Does that mean we have to live with war? It may well be that everyone of us has to adapt to the idea that in a not-so-remote future our town may be vaporized in a nuclear explosion or that you or your son will be asked to charge a machine gun nest armed only with a bayonet and all that, again, in the name of the war that will put an end to all wars. If war is a collective phenomenon that happens at the level of states and governments, then there is nothing you can do to avoid it, individually, or as a group. It is a meager consolation to know that this is the way the universe works.

Perhaps the best we can do, at this point, is to report the advice of a stoic philosopher, contemporary to Seneca, Epictetus, who in his *Enchiridion* ("The Manual") wrote that

"Some things are in our control and others not. Things in our control are opinion, pursuit, desire, aversion, and, in a word, whatever are our own actions. Things not in our control are body, property, reputation, command, and, in one word, whatever are not our actions. The things in our control are by nature free, unrestrained, unhindered; but those not in our control are weak, slavish, restrained, belonging to others. Remember, then, that if you suppose that things which are slavish by nature are also free, and that what belongs to others is your own, then you will be hindered. You will lament, you will be disturbed, and you will find fault both with gods and men. But if you suppose that only to be your own which is your own, and what belongs to others such as it really is, then no one will ever compel you or restrain you. Further, you will find fault with no one or accuse no one. You will do nothing against your will. No one will hurt you, you will have no enemies, and you not be harmed."

Famines, Epidemics, and Depopulation: The Zombie Apocalypse

Fig. 3.13 Bridget O'Donnell with her children, victims of the great famine that struck Ireland in 1845 (from Illustrated London News, December 22, 1849)

In 1968, George Romero directed a low-cost, black and white movie titled, *The Night of the Living Dead.* It was a great success that soon became a cult. Evidently, the film struck something deep in the human psyche with its theme of the dead rising from their tombs to devour the living. The movie critic Roger Ebert wrote about it that "I felt real terror in that neighborhood theater last Saturday afternoon" [65] and I have a personal recollection of having seen people vomiting in the hall of the theater after having watched the movie (Fig. 3.13).

The term "zombie" wasn't used in Romero's movie but it was the start of the genre that we call today "zombie apocalypse," plots involving a large number of 'undead' people haunting towns and suburbs in the search of live humans to kill and eat. But why this fascination with zombies in our times? How is it that we created a genre that never existed before in the history of human literature? Can you imagine Homer telling us that the city of Troy was besieged by zombies? Did Dante Alighieri find zombies in his visit to Hell? How about Shakespeare telling us of Henry V fighting zombies at Agincourt?

If something exists, there has to be a reason for it to exist and I think there is a reason why the zombie theme is so popular in our times. Literature always reflects the fears and the hopes of the culture that created it; sometimes very indirectly and in symbolic ways. And, here, it may well be that zombies reflect an unsaid fear present mainly in our subconscious: *starvation.*

Let's start with a typical feature of zombies: the black circles around the eyes. Zombies are supposed to be cadavers that somehow maintain a semblance of life. But do cadavers have this kind of eyes? Maybe, but the facial edema that creates the dark eye socket effect is also typical of malnourished people. If you look at how artists drew the starving Irish people during the Great Famine that started in 1845, they clearly perceived this detail. In the figure at the beginning of this section, you can see a rather well-known image of Bridget O'Donnell, one of the victims of the Great Irish Famine—note the darkened sockets of her eyes. Her children, too have the same dark circle around their eyes. Of course, comparing the starving Irish to zombies does not imply a lack of respect for the Irish men and women who perished in one of the greatest tragedies of modern times, but it tells us something about how starving people are perceived in our collective imagination. Zombies seem to be the perfect image of the effect of famine, not just in terms of their emaciated aspect but also in terms of their behavior.

Now, imagine that something happens that stops the supply of food to the aisles of your local supermarket. Imagine that it happens to all supermarkets in your region: maybe a shortage of fuel, maybe a war, maybe something else, it is anyway something that could happen [47]. People living in suburban areas would be first surprised, then angry, then desperate, and, finally, starving when their home stocks of food run out. Even before that, they would have run out of gas for their cars; the only system of transportation available to them.

Unless the government could (and would want to) intervene, the inhabitants of the suburbs would soon become emaciated, blundering, hungry people haunting the neighborhood and the shopping malls in the desperate search for something to eat. When they run out of canned food, some may turn to cannibalism, as zombies do in movies. Some may be able to put their hands on a good supply of guns and ammunition and then they could play king of the hill for a while, stealing most of the remaining food from those who hoarded it and shooting dead the poor wretches who still lumber in the streets, one more trope of zombie movies. The old Latin adage "*mors tua, vita mea*" becomes the rule. As Seneca Collapses go, this case is among the worst possible ones!

Of course, this is not a prediction and we can hope that nothing like that will ever happen, but it cannot be ruled out as impossible. I am not the only one to have noted this point, Terrence Rafferty wrote in 2011 in a literary review in The New York Times [66] that,

> ... it's a little disturbing to think that these nonhuman creatures, with their slack, gaping maws, might be serving as metaphors for actual people — undocumented immigrants, say, or the entire populations of developing nations — whose only offense, in most cases, is that their mouths and bellies demand to be filled.

Fictionalized catastrophes ("it is only a movie!") are surely less threatening than those that are described as likely to happen for real. It is a curious trait of the human mind but it may be that the only way for our mind to cope with possible catastrophes to come is to see them as fairy tales. But what are the chances of a real major famine striking the world of our times?

The general opinion on this point seems to be that famines are a thing of the past. You probably know the story of the wrong predictions made by Paul Ehrlich [67] with his 1968 book *The Population Bomb*, where he wrote that "In the 1970s hundreds of millions of people will starve to death." It was another example of how the secret for making wrong predictions consists of extrapolating the current trends. Indeed, the 1950s and 1960s had seen several large famines, including the Great Chinese Famine of 1959–1961 which caused at least 15 million deaths. So, the idea that famines were common and that they would continue in the future was a common perception in the 1960s. It may not be a coincidence that Ehrlich's book and the zombie movie by George Romero appeared in the same year.

On the other hand, if Ehrlich made a wrong prediction in terms of timing, that doesn't mean he was wrong in terms of substance. If he had framed his views in terms of a scenario rather than a prediction, then it would not be so easy to sneer at him, something which seems to have become a popular pastime. So, always remembering that the future is never like the past, what can we say about the possibility that major famines could cause local—or even global—collapse of the human population?

We know that there are more than 7.5 billion people alive on Earth today. Evidently, if they are alive, it means enough food is produced to keep them alive but that, of course, does not mean abundant food for everyone. Many people in poor countries are undernourished, while in rich countries many suffer from the opposite problem: obesity. That may, actually, be another form of undernourishment: it is known that poor people eat more "junk food" than the rich, and that they are also more overweight on the average [68]. A common interpretation is that the diet of poor people in rich countries lacks in vegetables, fish, and fruit and so it cannot provide the vitamins and micronutrients needed for good health. They try to compensate by eating too much, in particular in terms of carbohydrates. Even though the direct link between sugar and obesity is controversial [69] this interpretation can explain many features of the current obesity epidemics in the West, a multi-scale, systemic problem [70]. Surely, not something that can be explained by simply assuming Westerners are too rich.

But it is also true that nowhere in the world today do we see the kind of famines that occurred decades ago, with starving people stumbling around and looking like zombies before falling dead on the sidewalks. The lull in famines appears clear in the historical data for the past century or so [71]: there was a maximum of famine-related deaths in the 1940s, with more than 18 million deaths during the decade. In comparison, the decade of the 1980s had slightly more than 1.3 million deaths. The 21st century saw a certain increase with more than 2.8 million deaths during the 2000s, still much lower than the historically recorded maxima. These are not negligible numbers but they do indicate an improvement. Evidently, the world's food production system has been able to cope with the increasing world population, so far at least. By all means, it was a remarkable achievement (Fig. 3.14).

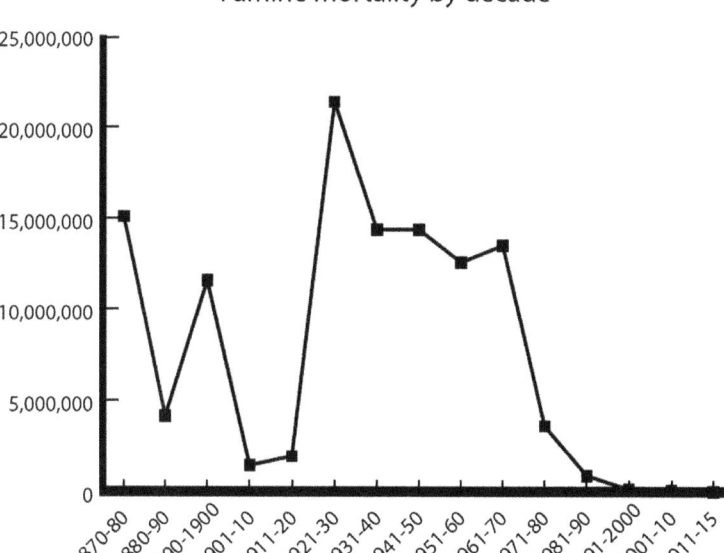

Fig. 3.14 Famine mortality in the world. Data from the World Peace Foundation (2015)

The decline in famines is normally attributed to technological factors. Fertilizers, pesticides, and mechanization greatly increased the yield of production per area unit, creating what we call today the "Green Revolution." The term gives the impression of some sudden technological improvement but that was not the case: yields gradually improved as the result of progressive innovation in cultivation techniques. But more than that, the disappearance of famine was due to container ships and low-cost trucks that made it possible to transport food everywhere in the world. In turn, these ships would not have transported food had they not been coupled with political and market-based measures. After World War II, providing food for the population of poor countries was seen as a way to avoid the diffusion of Communism and, also, as a simple way to subsidize the overproduction of Western agriculture [72]. That was one of the factors generating the economical and political system we call "Globalization." With the world having become one single giant market, anyone can use dollars to purchase food from everywhere and have it delivered to where they live. Since food is so cheap and since its purchase is often subsidized, the result has been a capillary distribution of food everywhere. Paul Ehrlich had not understood the importance of these factors when he predicted that hundreds of millions of people would starve to death. They haven't. Not yet, at least.

The problem is that, if there is enough food for 7.5 billion people today, that does not mean there will be enough in the future. It is another case of the main rule of prediction: the future is never like the past. So, you would be making the same kind of mistake Ehrlich made if you were to extrapolate the current situation and from that conclude that there will be no more famines in the word. The destruction of fertile soil, the depletion of aquifers, the increased reliance on depletable mineral fertilizers, to say nothing of climate change, are all factors that may make the future of food supply much harder than it is nowadays for humankind. The problems will be exacerbated if the population continues to grow.

Note also that the world's food supply system is a complex one that links technological, economical, and political factors. As we saw in this book, these systems are subjected to the kind of sharp crash that we call "Seneca Cliff." The slow growth of the system lulls you into a false sense of security until you find yourself falling down the cliff. So, famines are often accompanied by epidemics and wars. An undernourished population is easy prey of microbes in various forms and in ancient times famines and plagues went together or followed each other. Then, the stress generated by famines may generate political stresses which, in turn, generate violence. Conversely, wars may generate famines, sometimes intentionally provoked by one side to weaken the other. It is always the same mechanism that I dubbed the "Seneca Crunch:" all the negative factors gang up together to bring the system down.

Here are some examples of famine-related population collapses that took place in the past. First of all, here are the data for the Chinese Famine of 1959–1960s [71] (Fig. 3.15).

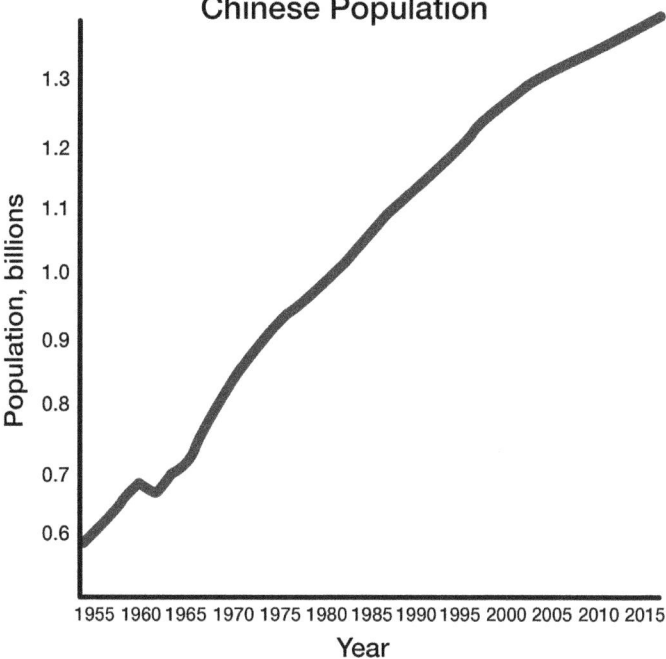

Fig. 3.15 Demographic data for China. Data from "Our World in Data" [71]

In terms of sheer numbers, with 15 million deaths directly or indirectly attributable to lack of food, it was one of the largest tragedies generated by famines in the historical record. Yet, note how these 15 million victims caused only a barely detectable dent in the Chinese population, about 2% of the total number, at that time close to 700 million people. The number of births rebounded just a few years after the famine phase and in practice, the trajectory of Chinese population growth was not significantly affected by the event.

Here is, instead, a graph of the victims of the Irish famine of 1845–1849. The rapid population drop was not caused just by starvation and the associated sicknesses, but also by emigration, but even that was a consequence of the lack of food. Losing some 2 million people in a few years, about one quarter of the total population, was not just a human tragedy but a social and cultural disaster that led Ireland, among other things, to lose its national language, Gaelic, to be replaced by English (Fig. 3.16).

Fig. 3.16 The Population of Ireland. Data from the Maddison Database www.ggdc. net/maddison/Historical_Statistics/horizontal-file_02-2010.xls

Finally, a third example where we see both phenomena at play in the same country, a transient loss of population and a long lasting one: Ukraine (Fig. 3.17).

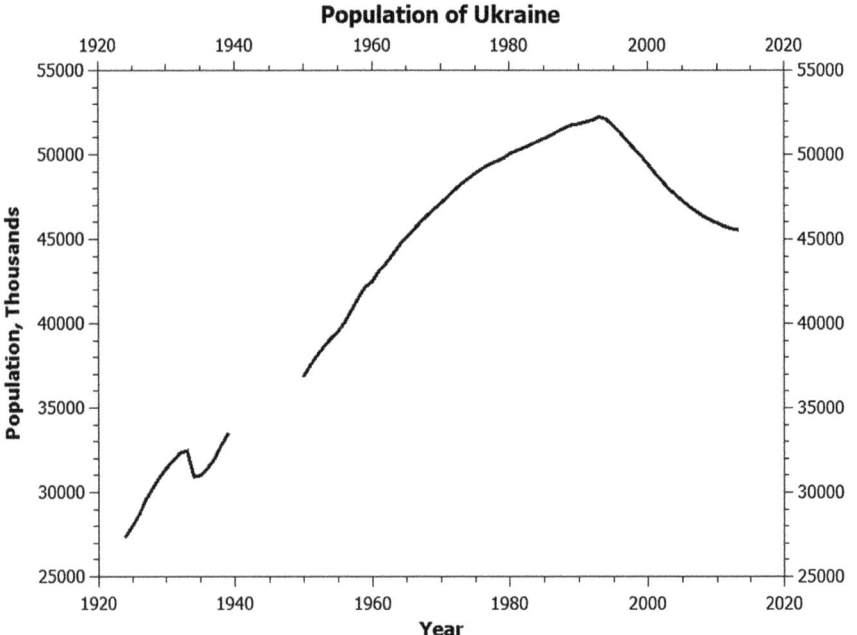

Fig. 3.17 The Population of Ukraine, including the effect of the Great Famine of the 1930s. Data from Wikipedia, https://en.wikipedia.org/wiki/Demographics_of_Ukraine

The data are incomplete but they clearly show two phases of population decline in Ukraine. The first corresponds to the Great Famine of 1932–1933 which affected not just Ukraine but large areas of the Soviet Union. It was a tragic famine with some 2 million deaths in Ukraine alone, perhaps more. But, tragic as it was, it is a transient in the population growth curve. The Ukrainian population may have suffered another decline phase during WW2, but the data are missing. In any case, in the 1950s, the population had rebounded and the growth phase that followed lasted until Ukraine reached its population peak at about 53 million, around 1990. Then, with the fall of the Soviet Union, in 1991, decline started, lasting to this day. This decline was not caused by famines, at least not the kind that lead people to die by starvation. But the quality of nutrition is likely to have declined together with the quality of health care and that has been increasing the death rate, especially with the elders. At the same time, Ukraine saw a reduction of birth rates in the same way as most former Soviet countries. We cannot say if the currently ongoing decline is irreversible, but it may well turn out to be.

These are just examples of modern famines, representing a phenomenon that has been common in history. Famines happen: sometimes they are transient phenomena generated by some natural disaster such as extended droughts, worsened by the mismanagement of corrupt or incompetent governments, or both. Sometimes they are systemic trends caused by the population having overcome the limits of what the local agriculture can sustain. This limit is not a fixed entity, it may be overcome by better agricultural technologies as well as by social and economic factors that favor better distribution of the food. The limit may also decline as the result of the depletion of the key resource for agriculture: fertile soil destroyed by overexploitation.

Whatever the case, in some historical examples it is clear that some limit was breached: countries such as Ireland in 1845 and Ukraine in 1991 were simply unable to sustain the population level they had reached. The return to sustainable limits took the shape of an apocalyptic disaster in Ireland, where the underdeveloped transportation and financial infrastructure of the country made it impossible to compensate for the collapse of the agricultural production in the South-Eastern regions. It was less dramatic in Ukraine, but it was still a major event. The case of Ukraine, as well as of several former Soviet countries, shows that there is no need of seeing people dropping dead in the streets for the population to decline. Apparently, young people tend to think that their children will have few opportunities in an economically declining system and abstain from procreating more than a few. The elder, then, must cope with poor nutrition and lack of health care: that may not kill them right away, but surely lowers their life expectancy. A similar effect is taking place in most Western European countries in terms of lower birth rates, but the life expectancy remains high and the reduced of the native population is compensated by immigration.

An often discussed interpretation of famines is that some of them are "man-made," that is are the result of specific evil actions carried out by governments and designed to starve and kill people. The best-known case is that of the Irish famine of the mid 19th century, said to be a crime perpetrated by the evil British government against their Irish subjects, but this accusation is heard for other modern famines. The Soviet government is blamed for the 1932 famine in Ukraine and the Chinese government for the Chinese famine of 1959. Now, it is true that governments are not benevolent associations, rather they tend to be among the most deadly organizations ever created by humankind. According to Rudolph Rummel [73], over the 20th century, some 256 million people were exterminated, directly or indirectly, by government actions in what Rummel calls "democides," a term that includes not only the victims of regular wars, but also other kinds of actions designed, for instance, to starve people to death.

Overall, though, it seems that governments are rarely interested in killing their own citizens: they need them as taxpayers or cannon fodder. On the contrary, they often try to multiply them: encouraging natality is a traditional policy of dictatorships. But governments do engage in the extermination of minorities, people who are identifiable and who can be labeled as enemies because of their race, language, religion, and ideology. For this purpose, they normally use conventional weapons: the problem with famines as weapons for ethnic cleansing is that one cannot easily distinguish friend from foe unless the population to be exterminated is localized in a specific geographical region.

For what I can say on this matter, I see no evidence that the British government willingly acted to create or worsen the Irish famine of 1845. They had no interest in killing a population that was providing a revenue for them. But it is true that they were slow and inefficient, and sometimes their actions worsened the situation. This is not surprising: another well-known characteristic of governments is that they are poor at managing complex systems. Other cases are less clear-cut but, personally, I tend to think that incompetence is normally a better explanation than evil intention for the great famines of history.

What about the future? Will we see new major famines in the world? A commonly heard question on this point is "how many people can the Earth support?" It is an ill-posed question for several reasons. It should be, rather, "how many people can the Earth support *indefinitely*." It is a truism that the

Earth can *now* support nearly 8 billion people: it is doing just that. But that is done in large part by "mining" a non-renewable resource: fertile soil. So, the large human population living today on the Earth may be just a transient phenomenon, way above the carrying capacity of the planet.

We often hear, today, about the "number of earths" we would need in order to provide for a long time the amount of resources we are consuming today. This is a concept related to that of "ecological footprint" proposed by Wackernagel [74]. Using the concept of footprint, we can calculate that, today, we are using almost 2 earths, and if everyone were to live at the same level of consumption of natural resources as the United States then we would need something like five Earth-like planets. That may force us to "return" well below the sustainability limits and that may turn out to be somewhat uncomfortable for most of us.

But there is a deeper reason why the question of the population limit is an ill-posed question. It is because famines and the related epidemics in history have always been localized in specific regions of the world. When disaster strikes, it is hard for a starving and sick population to move far away in search of food. In Ireland, for instance, people had no transportation other than their feet and most of the victims of starvation died close to their villages. In modern times, it is much easier to transport food where it is needed rather than transporting people where there is food available. As long as the economic system we call "globalization" remains active, this capability provides a remarkable resilience to the food production and distribution system. But things may change with the fashionable trend of building walls to mark state borders further limits the mobility of the poor and provides a barrier against the possibility of masses of hungry people swamping richer regions. That may result in large regions of the world experiencing disastrous famines, while others manage to maintain a sufficient food supply. It would be nothing different from the situation before globalization, when famines where a normal feature of life, everywhere.

Overall, famines may be one of the most clearly perceived threats nowadays, although it is a perception rarely expressed in the open. As individuals, we may want to prepare for a major famine by stocking supplies in the basements of our homes or by stockpiling guns and ammo in order to steal the supplies of our neighbors. It is doubtful (to say the least) that these strategies will be effective. If a major famine strikes, survival is possible only acting together as a whole society. Whether this will be possible in the world we call the "West" which puts so much emphasis on individual reliance, is all to be seen.

The Big One: Societal Collapse

Fig. 3.18 The giant stainless steel monument to the Soviet Worker and the Kolkhoz Woman in Moscow. It was created by Vera Mukhina in 1937 to symbolize the march forward of the then recently created Soviet Union (1922–1991). The Union was the last (so far) of the long series of empires that ebbed and flowed along human history (Image by Limitchick https://en.wikipedia.org/wiki/Worker_and_Kolkhoz_Woman#/media/File:The_Worker_and_Kolkhoz_Woman.jpg)

In 1992, I received an email from Russia. Written in very good English, it contained wishes for my birthday and a proposal of research collaboration. It arrived from a research institute in Moscow where some Russian physicists had been working in the field of science where I was active in at that time, surface science. With the Soviet Union having disappeared just one year before, they were looking for international contacts and collaborations. Without their salary, and without funding for their research, the researchers of former Soviet countries were being forced to find jobs as janitors, clerks, or translators, while many of them had to leave Eastern Europe to continue their career in the West. That was the start of my involvement with former Soviet researchers and research institutions, especially in Russia and Ukraine (Fig. 3.18).

Witnessing the effects of the Soviet collapse from inside was a sobering experience and it made me wonder about the reasons that had brought down the Soviet Union. At the time, I tended to agree with the generally accepted explanation that Francis Fukuyama had termed the "End of History" [75]. In this view, the crash had been due to the inefficiency of the Soviet State and it had demonstrated the superiority of the Western Political system.

But the more I understood Russia the more I became dubious about this optimistic interpretation. With all its defects, its quirks, its ideological bent, its overblown bureaucracy, and its many more problems, the Soviet Union was still a state that encompassed a large part of Eurasia and nearly 300 million people. Its scientific achievements had been remarkable and had included the first artificial satellite, the first man in space, and mounting a serious challenge to the West in the race to the Moon of the 1960s. To say nothing about having defeated the German invasion during WW2 at a loss of more than 20 million soldiers. If you ever took a train of the Moscow subway and saw the elaborately decorated stations there, you could not miss the fact that the Soviet Union had been much more than just a dictatorship kept together by its secret police. And, although the research work of the Soviet scientists was mostly unknown to their Western counterparts, it was often at the same level, if not better.

Mostly, it was the resilience of the Russian people that impressed me. I still remember a scene that I witnessed in the 1990s, probably at the darkest moment of the economic crisis in Russia. At that time, the local currency, the ruble, had become nearly worthless and most transactions were made in dollars, even for ordinary items such as food in the supermarkets. So, at the exit of a train station in Moscow, I saw maybe a dozen Russians, men and women, lined up along the sidewalk, each one with something in their hands: a shirt, a pair of shoes, a hat, or other everyday items. They were selling what they had for a few rubles. At first, I thought that they were doing that out of desperation. But then I thought it better: these people were not desperate, they were making a statement. They were sharing what they had with the others and, in doing so, they were saying that rubles were still money and that Russia was still an independent country with its national currency. Eventually, they were vindicated: over the years, Russia returned to use rubles and the economy rebounded to a degree of reasonable prosperity. Many scientific institutions in Russia and in the former Soviet Union have returned to their previous level of excellency and I am glad to have been able to give a hand in the task although, of course, the merit goes entirely to the obstinacy, the persistence, and the hard work of the Russian researchers.

My experience with the Russian collapse went in parallel with that of Dmitry Orlov, an American of Russian origin who also personally experienced the effects of the collapse of the Soviet Union. Orlov reported his experience and his ideas in a series of books, the first one (2011) with the title *Reinventing Collapse. The Soviet Example and American Prospects* [76]. This title, I think, explains what the book is about. Orlov can speak native Russian and his knowledge of the Russian society is obviously much better than mine, but his experience agrees very much with mine. The collapse of the Soviet Union was not a simple question of a wrong ideology put to rest: it was due to deep reasons that had weakened the Soviet society from inside causing it to follow a trajectory that inevitably led it to decline and disappear. According to Orlov, the same factors are at work to bring the Western society to an inevitable future collapse.

I think that if Tolstoy had witnessed the collapse of the Soviet Union, he would have interpreted it in the same way as he had interpreted the invasion of Russia by Napoleon's armies. "It happened because it had to happen." It did not and it could not happen because some puffy leader had decided something. In other words, it had little or nothing to do with the often-heard story that Ronald Reagan, Margaret Thatcher, and King Fahd of Saudi Arabia together had managed to bring down oil prices in order to lower the revenues that the Soviet state obtained from oil export and make it collapse. The collapse of the Soviet Union had a lot to do with crude oil, but certainly not in terms of conspiracy theory. Nor the fall could be caused by the Soviet leader of the time, "Mad Misha" Gorbachev, alone, who was so naive to be easily cheated by the promises of the evil Western leaders.

The Soviet Union went down and disappeared from history just as one more example of how states, empires, and entire civilizations collapse. So far, no human civilization has survived this destiny, none has lasted more than a few thousand years without undergoing at least some kind of collapse, maybe re-emerging stronger afterward, but also deeply changed. Some civilization were smashed by external events: the Minoan one on the shores of the Mediterranean sea was probably destroyed by the mega-eruption of the Thera volcano during the mid-second millennium BCE. Some civilizations were destroyed by the military power of technologically more advanced ones, such as the Aztec and the Inca Empires, destroyed by the Spanish armies during the 16th century CE. But, in the great majority of known cases in history, civilizations and empires fell by themselves or, if defeated by foreign powers, because they had been greatly weakened for internal reasons. Between 1934 and 1961, British historian Arnold Toynbee (1889–1975) wrote *A Study of*

History describing the rise and fall of the 23 civilizations he had studied. His conclusion was that "civilizations die from suicide, not by murder." He had identified a typical feature of complex systems, tending to collapse because of the sometimes deadly mechanism of reinforcing feedbacks. That was certainly the case of the Soviet Union, neither militarily defeated nor hit by an asteroid: it collapsed mainly for internal reasons.

The collapse of civilizations is one of the most controversial subject of historical study. There are, literally, hundreds of different explanations for some of the most spectacular falls, such as in the case of the Roman Empire. It seems that these explanations appear and disappear in reason of the current worries of our own civilization. For instance, historian Kyle Harper recently transferred to the ancient Roman Empire one of our major worries: climate change, arguing that it was at least one of the major causes of the fall [77]. That involves stretching the data a little, to say the least, since the data show no evidence of significant climatic changes in Europe until well after that the Roman Empire was in its death throes [78].

In reality, the historical cycles of empires and civilizations indicates that there have to be generally valid mechanisms that bring about their fall. In recent times, a certain agreement seems to be emerging on this point and a pioneer in this field has been Joseph Tainter with his idea of the "diminishing returns of complexity" [79]. According to Tainter, civilizations tend to expand and, as they do, they develop internal structures that are used to cope with external and internal threats and challenges: the army, the legal system, the police, the bureaucracy, and many others. Tainter's idea is that the efficiency of these structures diminishes as they grow larger. That is, they become less and less effective at performing the tasks they were built for. According to Tainter, this phenomenon leads eventually to diminishing returns. This is the mechanism that brings down the stupendous structures we call "civilizations" or "empires."

Tainter's ideas are steeped in the science of complex systems, but they are qualitative. Actually, Tainter does support his interpretation with archaeological and historical data but only indirectly and one question that remains unresolved is how exactly diminishing returns bring collapse rather than just slow down growth.

Recently, together with my coworkers Ilaria Perissi and Sara Falsini, we tried to reproduce Tainter's ideas using a model developed using the tools of system dynamics [80]. We found that the basic concept proposed by Tainter, diminishing returns, can be reproduced by the model. But we also found that it is not just the increase in size that reduces the efficiency of the structures of society: it is the combined effect of the higher cost of natural resources and that of having to fight pollution. When these effects are taken into account, the model produces a curve for the diminishing returns of complexity that looks qualitatively similar to the one proposed by Tainter (Fig. 3.19).

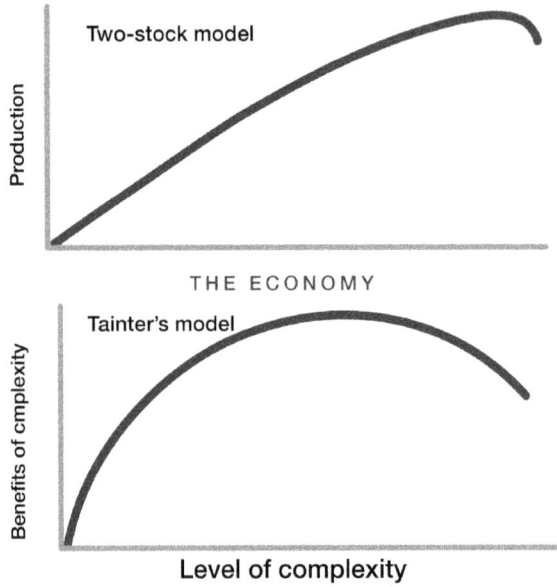

Fig. 3.19 The main results of the Study on Civilization Collapse performed by Bardi, Falsini, and Perissi in 2019 compared with Tainter's curve. In the study, we assumed that the level of complexity of a civilization is proportional to the size of its economy [80]

Most civilizations in history seem to arise from the availability of some abundant and cheap natural resource. The Roman Empire grew on the production of precious metals from gold and silver mines, in particular those of Northern Spain. Our current world empire has grown on the availability of abundant and cheap fossil fuels, first coal, then—currently—crude oil. But we have seen how natural resources tend to be overexploited and also how this phenomenon leads to their rapid depletion and, often, to a rapid crash of the system: it is the basic mechanism of the Seneca Collapse.

The Soviet Union was an empire mostly based on its vast mineral resources, and it was unable to escape the fate of other mineral-based empires: collapse caused by overexploitation. The fall of the Soviet Union was amply predictable much before it happened and it was, indeed, predicted by Soviet researchers themselves. On this point, Dennis Meadows, one of the main authors of the 1972 study *The Limits to Growth*, gave a talk in Moscow in 2012, telling how Soviet researchers had applied the same models to study the economy of the Soviet Union, finding that the system would soon collapse. They published their results in 1980 in a book (in Russian) titled *The Soviet Union and Russia in the global system*. According to Meadows, in the 1980s, Viktor Gelovani, first author of the Russian book went to the leadership of the country and he said, 'my forecast shows that you don't have any possibility. You have to change your policies.' And the leader said, 'no, we have another possibility: you can change your forecast'.

The 2012 talk by Meadows has disappeared from the Web, but its main points are summarized in an article of mine on the blog *Cassandra's Legacy* [81]. Meadows' statements are confirmed by the work of Eglé Rindzevičiūtė who wrote an excellent article that tells the whole story [82]. It is clear that several Soviet scientists knew very well the "Limits to Growth" story and its methods and results, even though the study was officially rejected by the Soviet Government as the result of decadent Western science. These Russian scientists understood that the same factors that the study had considered for the whole world would apply to the Soviet Union. They seem to have made a considerable effort to warn the Union's leadership that the system was going to collapse. The reaction of the Soviet leadership was the same as it was in the West: both Soviet and Western leaders were completely tied to the concept of "growth at all costs" and refractory to changes. So, the warning was ignored and, as usual, ruin followed. It may well be that the straw that broke the back of the Soviet camel was the increasing costs of oil production, as argued by Douglas Reynolds in his book *Cold War Energy* (2016) [83].

So, what can we expect from the future? Are we going to see the Western Civilization following the same path as the old Soviet Union? It is perfectly possible that many of the readers of this book will experience this kind of future. So, it may be worthwhile to listen to the forecast of someone who experienced the Soviet Collapse: Dmitri Orlov. In his book, *The Five Stages of Collapse*, [84] he summarizes how the collapse of a complex society takes place.

Stage 1: Financial collapse
Stage 2: Commercial collapse
Stage 3: Political collapse
Stage 4: Social collapse
Stage 5: Cultural collapse

It may well be that we are already experiencing the early stages of the process, mainly in the form of financial troubles. The financial shock of 2008 was somehow remedied by what was called "quantitative easing" (QE) and consisted mainly in pumping large amounts of currency into the system, It seems to have worked, for a while at least, but many economies in the world have not completely recovered and maybe never will.

The problem with using financial tools to solve the crisis is that you can have all the virtual money you want, but people cannot eat virtual food, nor power their cars and homes with virtual energy. This is a problem that the ancient Soviet Union already had with the ruble, which gradually became a worthless currency and gave rise to the well-known joke that said, "they pretend to pay us and we pretend to work for them." In our world, money in the form of dollars is valuable even if it is fully virtual as long as you can exchange it for oil and all the products made from oil, from clothes to food, to fuel for your car. If (when) oil ceases to be available on the world market, then all the dollars in the world will become worthless.

Indeed, the 2008 financial collapse was directly related to the spike in oil prices which had reached the record value of $150 per barrel that year—in turn related to the high costs of extraction caused by depletion. The tumultuous arrival of shale oil on the market gave us at least a decade of pause, with oil prices remaining high on the average, but never again reaching their 2008 values. From where we stand now, everything is possible: we can see more instabilities, the collapse of the shale oil industry, and more perturbations of the fragile oil production and supply system which might well bring down the whole financial market, this time in a way that no new quantitative easing trick will fix. In that case, we would see nothing less than a stroke for the whole system, a true Seneca cliff of the worst kind.

Following the financial collapse, we might see the three other Seneca Horsemen of the Apocalypse: commercial, political, and social collapse. We are not there, yet, but if the whole system loses the fundamental communication ingredient that keeps it together: money, it means that people will still have things to sell and there will be people wanting to buy them. But, without money, not only buyers cannot pay sellers, but the goods cannot be delivered. It means that the shops run out of everything. Will you run out of food and starve? Maybe. Already in 2008, a consequence of the financial collapse was that the ships carrying merchandise all over the world stopped moving. It did not last long enough to cause the death by starvation of billions of people, fortunately, but that could be the consequence of a longer-lasting financial shock.

Some evident symptoms of commercial collapse are already visible all over the West. If you live in a poor area of your country, you may have noticed that your options in terms of shops and merchandise available have been drastically reduced. Then, of course, you may buy whatever you want on Amazon.com, but only if the financial system still lets you do that and if there is a still functioning delivering system taking it to your door. On a larger scale, the numerous economic sanctions enacted by the US government and their allies against countries perceived as enemies prefigure the breakdown of globalization as a worldwide commercial system.

Political collapse goes together with commercial collapse. Without money, people cannot buy anything and risk starving or freezing (or both). At this point, the only possibility to keep the social fabric together is for the government to intervene and provide emergency supplies, as they normally try to do in the case of large natural disasters. But the historical record on governments managing catastrophes is not good. Will they really want to help people? Or will they rather save themselves and their cronies?

Social collapse also comes together with commercial and political collapse. People will do what they can to help each other, but if things really go out of control the result may be true mayhem. We are already seeing evident symptoms of the breakdown of the social fabric in the West in the increased political polarization. In a two-party system, people try to elect people holding ideas similar to theirs, but normally the people voting for the other party are not supposed to be monsters to be hated, as it seems to be the rule in our times. The kind of ideological hate that pervades our society nowadays is a true fracture of the social fabric. Racism, hate for foreigners, defensive walls, every man for himself, bomb them back to stone age, gun and ammunition in everyone's basement, and more. So far, a veneer of civilization seems to be still holding, but never forget that someone said that the only thing that separates civilization from barbarism is two hot meals. Is it a prophecy? No. It is a scenario. And scenarios sometimes come true.

There remains the final stage, cultural collapse, the phase in which people cease to recognize themselves in the culture that supported the state that collapsed. The Romans stopped being pagan and the Russians stopped being communists. Neither was necessarily a bad thing, it was part of the unavoidable force that moves complex systems: change. Passing the tipping point that is collapse, the system needs to re-adapt. It did so for past collapses, it will do that for the future one, at least if we cannot manage to mitigate it.

Cultural collapse is a major change, it is actually gigantic. Think of what happened to the Roman Empire: it reverted to the political organization that had existed before, city-states and chiefdoms. But it was not just a return to the past: it was a radical change in many things. The Western Roman Empire left as an inheritance its imperial language, Latin, which became the sacred language of the Catholic Church. Latin was the governance tool in what became a social experiment never attempted before in the world: the Middle Ages mimicked the old imperial order but, instead of money, it used the spiritual benefits that the Church dispensed to the believers.

I do not mean that the fall of the modern Western Empire will bring back the Catholic Church, even though you never know what to expect from an organization resilient enough to have been able to survive for at least

1500 years. What I mean is that the cultural change that awaits the West will be enormous and radical. It may bring humankind to a new stage of social organization by going in parallel with the evolution of the human brain that led us to the axial age in just a couple of millennia. If we manage to maintain some of the technological capabilities that our tumultuous times have developed, we may one day emerge into a new civilization that might be benevolent and merciful to itself and also toward all the creatures of this planet.

Apocalypse: The Collapse of the Earth's Ecosystem

Fig. 3.20 An interpretation of the four horsemen of the apocalypse by Arnold Bocklin (1927–1901) in an 1896 painting titled Der Krieg (the War). Apocalypse means "revelation" in Greek, but it is commonly understood as referring to the end of the world

Imagine you are living in Jerusalem in the year 70 CE. And imagine that you have a chance to climb on one of the ramparts, on the walls, and take a look at what is happening outside. Out there, you see the encampments of four Roman legions surrounding the city in full war posture, equipped with giant siege machines. At that point, you might be justified if you were to feel a certain sensation that the city is doomed.

Indeed, some of your fellow Jerusalem citizens seem to have become a little catastrophistic in their feeling. One is Jesus son of Ananias (Yeshua ben Hananiah), whose last deeds are so reported by Josephus in his "*The Jewish War*" written some years after the war.

> … he every day uttered these lamentable words, as if it were his premeditated vow: "Woe, woe to Jerusalem." Nor did he give ill words to any of those that beat him every day, nor good words to those that gave him food: but this was his reply to all men; and indeed no other than a melancholy presage of what was to come… Until the very time that he saw his presage in earnest fulfilled in our siege; when it ceased. For as he was going round upon the wall, he cried out with his utmost force, "Woe, woe to the city again, and to the people, and to the holy house." And just as he added at the last, "Woe, woe to myself also," there came a stone out of one of the engines, and smote him, and killed him immediately. And as he was uttering the very same presages he gave up the ghost.

Prophets of doom seem to be common in history whenever the situation starts looking hopeless for one reason or another. The list is long, with Yeshua ben Hananiah being just one of them. They are not usually seen with sympathy and their litanies are scoffed at. They may be compared to Chicken Little who thought the sky was falling because a nut fell on his head. But there seems to exist a basic phenomenon in human social groups that makes prophets of doom appear whenever there is a chance of some major disaster to occur.

As you surely noted, in our times doom-mongering has become a small cottage industry. A good example is the story of the planet Nibiru (or maybe Planet X, or maybe Herculobus, or whatever), said to have been aiming toward the Earth and scheduled to hit it in 2012, a prediction based—it seems—on an ancient Mayan calendar. Maybe the Mayans had ended their calendar with the year corresponding to our 2012 just because they had reached the end of the stone wall where they were engraving dates. In any case, the story became popular even though, of course, nothing larger than ordinary meteorites hit the Earth in 2012. The most recent version predicted that the planet Nibiru would hit the Earth in 2017. It was wrong, too, and it is possible that the arrival of Nibiru will be postponed to some future date.

Nibiru is part of a wave of imagined threats periodically sweeping the Internet in various forms and in various degrees of silliness. Some seem to be the domain of complete nuts: one is the "chemtrails" story that sees the innocuous trails left by aircraft as harmful chemicals spread by the powers that be in order to poison us. Other legends have a certain scientific basis although the threat may be wildly exaggerated, such as when some people fear that burning fossil fuels would consume the oxygen we breathe. It is true that we can measure a slight reduction of the oxygen concentration in the atmosphere, but it is minuscule and even burning all the known fossil fuel reserves would not lead to a decline large enough to affect human health.

Overall, existential threats seem to have a certain sales power. For instance, Listerine was marketed in the 1920s as a remedy against halitosis, or simply bad breath, that the creators of the advertising campaign aggressively described as a serious threat for people's social success [85]. Peddling Listerine as a way for girls to get a husband surely was a little aggressive, although not so bad as trying to scare people about the threat of a whole planet falling onto us. But the problem is that *some* prophets of doom turn out to have been right when the catastrophe arrives. After all, poor Yeshua ben Hananiah had correctly predicted the fall of Jerusalem in 70 CE. So, not all prophets of doom can be simply discounted as rambling madmen.

In our times, we surely face a number of threats large enough to be a source of worries not just for madmen and prophets, but for every one of us. For instance, every few years a group of thousands of the world's best scientists in climate and ecosystem matters get together to prepare a new report of the organization called IPCC (intergovernmental national panel on climate change). And, every few years, they tell us that if we do not stop burning fossil fuels, and fast, humankind is in dire trouble. What we are facing is the possibility of a disaster beyond anything ever experienced by humankind. The world's ecosystem is on track toward a temperature increase of about 3–4 °C over the next several decades, unless truly draconian measures are taken to reduce the emissions of carbon dioxide. And there is no guarantee that the warming would be limited to that: nonlinear feedback effects could increase it by 6–8°, perhaps even more. This level of warming would have an enormous impact on the ecosphere, threatening to destroy civilization as we know it, if not to cause the extinction of the human species. Now, if that is not apocalyptic I don't know what is. And we are not told about that by a screaming madman, but by the community of the best scientists in the world.

Facing the entity of the climate threat, the response of the human community has been weak, to say the least. You are told that you can fight climate change by such things as separating your waste, using low consumption light bulbs, buying local groceries, cycling, and other actions that seem to be conceived mainly to assuage one's guilty feelings, but little more than that. Most people tend to ignore the climate threat, while a small minority vocally maintains that it is all a hoax invented by a group of evil scientists who thought they could get more research grants and more graduate students by hyping a non-existent threat. The opinion polls show that the general opinion on climate change remains stuck at a 50/50 level with the public, that is about half of the people think it exists and is a serious threat, the other half think it does not exist or is not a problem. Recently, a survey carried out by Yale University [86] showed a certain movement toward a larger fraction of the public identifying climate change as something to be worried about. Maybe they are by now a majority, but it remains to be seen how many of them will be willing to pay money or make sacrifices in order to combat climate change. On this point, it is worth remembering that the "Yellow Vests" movement in France started in 2018 mainly as a result of fighting increasing fuel prices.

But for how long can people remain indifferent to the threat at the door of their cities? As the intensity of the threat mounts, it becomes more and more difficult to ignore. The change from indifference to terror may take the shape of a true tipping point, according to John Schlesinger'sassessment that "people have only two modes of operation, complacency and panic." The switch to panic may start small and there is evidence that it is, indeed, starting.

The accumulating knowledge about the phenomenon called "climate change" is indeed giving rise to at least one group of prophets of doom who claim that the end of the world is coming (or, as flea prophets would say, "the end of the dog is coming" as we can read in a *Far Side* comic by Gary Larsen). They tend to use the term "Near-Term Human Extinction" (NTE or NTHE) and one rather well known member of the group is Guy McPherson who keeps a blog titled "Nature Bats Last" [87]. NTE is not a monolithic concept, especially in the meaning of "near-term" but, according to McPherson, humankind could be already mostly or wholly gone by 2030, which is given as the last year for humankind on Earth. In a recent interview [87], McPherson stated that (emphasis in the original)

> *Specifically, I predict that there will be no humans on Earth by 2026*, based on projections of near-term planetary temperature rise and the demise of myriad species that support our own existence.

A rather bold prediction, to say the least. For the human population to go from nearly 8 billion to zero in seven years would be some kind of a Seneca cliff! Indeed, the NTE idea is normally discounted as the product of deranged minds. It must be said, in addition, that the members of the "NTE movement" do little to endear themselves to non-believers. They are often aggressive in the debate and tend to take a rigid attitude: NTE will happen because it has to. It is rather typical of groups embracing extreme, non-mainstream views. Being a tiny minority surely requires developing some defensive communication techniques. But the real problem with these prophecies of doom is that they encourage passivity. If we must die, why bother doing anything that could perhaps avoid it? One might as well take a vacation to Hawai'i as long as it is still possible. It might be worse if the NTE meme arrives to infest the minds of opinion leaders and of policymakers. In this case, if panic sets in, the response of the powers that be could be reckless, to say the least. It they were to come to the conclusion that climate change is caused by too many human beings, they could well decide that getting rid of most of them is a good idea. It is a disturbing idea, but we know how often and how easy in history entire societies tend to go into "extermination mode." It happened in the past, it can happen again.

In the end, is there a chance that the NTE believers might be right? Here, unfortunately, it is not possible to demonstrate that they are wrong. Yes, we can say it is unlikely, we can say that the models do not predict anything like that, that some extreme catastrophes such as the Venus effect seem to be ruled out by the physics of the Earth system [88]. But it is also true that climate-related catastrophes did take place in the Earth's past and we know that the results were mass extinctions, in some cases involving the extermination of most vertebrates. They are the results of massive volcanic eruptions known as "Large Igneous Provinces" (LIP). The effects of the largest LIPs on the biosphere was devastating [89] and it is now believed that the extinction of the non-avian dinosaurs was not—at least not directly—the result of the impact of a large asteroid but of a LIP that appeared in the region called today Deccan, in India. The End-Permian extinction was caused by another massive LIP appearing in the region called today Siberia. It wiped out about 95 percent of all vertebrate species on the planet [90].

The destructive effects of large igneous provinces are not directly caused by the heat generated but by the emission of large amounts of carbon dioxide (CO_2) in the atmosphere. As it is typical of complex systems, this forcing generates a cascade of enhancing feedback effects, including the release of methane stored in the permafrost and perhaps the combustion of coal deposits invested by the hot magma. The result is that the Earth is pushed on the other side of a tipping point into the condition described as "hothouse Earth" [91], as opposed to the conditions in which humans are accustomed to live, an "interglacial Earth." A "hothouse Earth," is a very hot Earth where temperatures are so high that large areas of the planets are uninhabitable by humans and possibly by most vertebrates, while mass extinctions occur as the result of factors such as the reduction in oxygen concentration (anoxia), the release of poisonous hydrogen sulfide from bacteria and more bad effects on life.

Now you can see what we are discussing about: a major kind of Seneca collapse not just for humans but for the whole biosphere. Of course, there is no active LIP on Earth, today, but what we are doing with our habit of burning what we call "fossil fuels" is having a similar effect: we are pumping large amounts of greenhouse gases into the atmosphere. The result is a forcing that could generate a cascade of feedbacks of the same kind of those generated by the ancient LIPs that destroyed most of the ecosystems of the time. As a further damning factor, today solar irradiation is stronger than it was during the past. It increases by about 10% every billion years and today it is significantly higher than it was during the largest mass extinction episodes of the past. It means that a smaller forcing is necessary in order to generate another major hothouse episode. No wonder that we seem to have entered the "sixth mass extinction" era [92]. The first five were caused by LIPs, the current one is human-made.

So, what are we facing, exactly? The climate models we use cannot provide an exact assessment of the effects of the reinforcing feedback loops that might lead to a climate tipping point, but there is a general agreement among scientists that some kind of "climate tipping point" exists [93], although nobody can determine its parameters exactly. The emphasis given in the Paris treaty about the need to stay below a maximum of 1.5 or 2 °C of warming is because of fears that going above these temperatures would mean passing the tipping point. But, again, these values have not been determined by quantitative calculations—they are a best guess, and they could be an optimistic guess.

Overall, we cannoty exclude that we are doomed, but it is also true that it is far from being sure and, for what we know, there is still plenty of room for maneuvering and, possibly, avoiding the worst. One thing that is reasonably certain is that the damage will be huge, well before hothouse Earth wipes humankind out—if it ever does that. Climate-related droughts may destroy a sufficiently large fraction of the agricultural production to cause widespread famines. Or the opposite phenomenon, floods, may do the same by washing out the fertile soil. Sea level rise may also cause a similar effect: making ports inoperable would interrupt the vital flow of food carried by container ships. It is not clear whether major weather phenomena, hurricanes or tornadoes, could have disastrous effects of the same magnitude, but that cannot be discounted. Facing these increasingly grave threats, humans could react in different ways: the basic rule of politics is to find a way to blame someone else, so a possible result would be to double down and increase the effort to ignore the threats. Or, conversely, a tipping point in perception could lead the elites to decide to move to desperate attempts to redress the situation by using geo-engineering, with all the unknowns involved. Who knows? It might even work. Or, the elites could decide to dump the poor and save themselves by occupying regions in the high north, or in the mountains.

Overall, for those of us who are not part of the elite, the future does not seem to be bright in terms of what climate change is bringing to us, and even if you happen to be part of the elite, the future looks hard as well. But the beauty of the future is that it cannot be predicted. So, we march into the future always equipped with an indispensable tool: hope.

References

1. Griffith, A.A.: The phenomena of rupture and flow in solids. Philos. Trans. R. Soc. London, A **221**, 163–198 (1921)
2. Odds of dying—data details—injury facts. Injury Facts. https://injuryfacts.nsc.org/all-injuries/preventable-death-overview/odds-of-dying/data-details/.   Last accessed 7 Mar 2019
3. Ponte Morandi di Genova, attacco degli Illuminati Sionisti sulla A-10. http://passaparoladesso.blogspot.com/2018/08/ponte-morandi-di-genova-attacco-degli.html. Last accessed 28 Feb 2019
4. Carvajal, A.M., Vera, R., Corvo, F., Castañeda, A.: Diagnosis and rehabilitation of real reinforced concrete structures in coastal areas. Corros. Eng. Sci. Technol. **47**, 70–77 (2012)
5. The crumbling history of Puerto Rico. Globe Gazette (2017). https://globegazette.com/news/world/photos-the-crumbling-history-of-puerto-rico/collection_ef289a27-d320-5743-9c31-31c9a70875d5.html. Last accessed 9 Jan 2019
6. Lugo, A.E.: Social-Ecological-Technological Effects of Hurricane María on Puerto Rico. Springer, Berlin (2019). https://doi.org/10.1007/978-3-030-02387-4
7. A crash course in probability. The Economist (2015). https://www.economist.com/gulliver/2015/01/29/a-crash-course-in-probability. Last accessed 27 Aug 2019

8. McGee, B.: Contracts of carriage: deciphering murky airline rules. USA Today (2017). https://eu.usatoday.com/story/travel/columnist/mcgee/2017/07/12/airline-contract-carriage/469916001/. Last accessed 4 Apr 2019

9. Glanz, J., Kaplan, T., Nicas, J.: In Ethiopia crash, faulty sensor on Boeing 737 max is suspected. The New York Times (2019). https://www.nytimes.com/2019/03/29/business/boeing-737-max-crash.html. Last accessed 30 Mar 2019

10. Satell, G.: A look back at why blockbuster really failed and why it didn't have to. Forbes (2014). https://www.forbes.com/sites/gregsatell/2014/09/05/a-look-back-at-why-blockbuster-really-failed-and-why-it-didnt-have-to/#6df219961d64. Last accessed 8 Mar 2019

11. Graeber, D.: Debt: the first 5,000 years. Melville House (2011)

12. Mitchell-Innes, A.: The credit theory of money. Bank. Law J. **14**, 151–168 (1914)

13. Kramer, S.N.: The Sumerians: their history, culture, and character. University of Chicago Press (1963)

14. Lapis scandali, M.L.: La pietra dello scandalo. Rome and Art. Available at http://www.romeandart.eu/it/arte-lapis-scandali.html. Accessed 8 Mar 2019 (2015)

15. Wagner, T.E.: Five reasons 8 out of 10 businesses fail. Forbes (2013). https://www.forbes.com/sites/ericwagner/2013/09/12/five-reasons-8-out-of-10-businesses-fail/#731dc61f6978. Last accessed 8 Mar 2019

16. Brown, E.: Student debt slavery: time to level the playing field. Counterpunch (2018). https://www.counterpunch.org/2018/01/05/student-debt-slavery-time-to-level-the-playing-field/. Last accessed 16 Apr 2019

17. Rhode, S.: Debt related PTSD and financial PTSD quietly hurts many. Get Out of Debt (2013). https://getoutofdebt.org/48667/financial-problems-and-debt-can-cause-ptsd. Last accessed 16 Apr 2019

18. Bardi, U.: The dark side of the internet: the "quantum code" scam and its implications. Cassandra's Legacy (2017). https://cassandralegacy.blogspot.com/2017/04/the-dark-side-of-internet-quantum-code.html. Last accessed 8 Mar 2019

19. Herley, C.: Why do Nigerian scammers say they are from Nigeria? In: Workshop on the Economics of Information Security (WEIS) (2012)

20. Taleb, N.: The black swan. Random House (2007)

21. McWhinney, J.: Massive hedge fund failures. Investopedia (2011). https://www.investopedia.com/articles/mutualfund/05/hedgefundfailure.asp. Last accessed 8 Mar 2019

22. Kruger, J., Dunning, D.: Unskilled and unaware of it: how difficulties in recognizing one's own incompetence lead to inflated self-assessments. J. Pers. Soc. Psychol. **77**, 1121–1134 (1999)

23. Bardi, U.: The sinking of the E-Cat. Cassandra's Legacy (2012). http://cassandralegacy.blogspot.it/2012/03/sinking-of-e-cat.html. Last accessed 19 Nov 2016

24. University of Bologna statement on e-cat testing. e-catworld (2012). https://e-catworld.com/2012/01/25/university-of-bologna-statement-on-status-of-e-cat-testing/. Last accessed 11 Jan 2019

25. Krivit, S.B.: Rossi's NASA test fails to launch. New Energy Times (2012). http://newenergytimes.com/v2/news/2012/Report-4-Rossis-NASA-Test-Fails-to-Launch.shtml. Last accessed 20 Apr 2019

26. Minsky, H.P.: The financial instability hypothesis, working paper (1992). http://www.levyinstitute.org/pubs/wp74.pdf

27. Doctorow, C.: Down and out in the magic kingdom. Tor (2003)

28. Townsend, S.: The naked environmentalist. Futerra (2013)

29. Meissner, M.: China's social credit system, China Monitor, May, 24, 2017. https://www.merics.org/sites/default/files/2017-09/China%20Monitor_39_SOCS_EN.pdf

30. Doctorow, C.: Wealth inequality is even worse in reputation economies. Locus Magazine (2013). https://locusmag.com/2016/03/cory-doctorow-wealth-inequality-is-even-worse-in-reputation-economies/. Last accessed 28 Feb 2019

31. Bullying and suicide (2017). http://www.bullyingstatistics.org/content/bullying-and-suicide.html. Last accessed 11 Jan 2019

32. Tehranirad, B., et al.: Far-field tsunami impact in the North Atlantic basin from large scale flank collapses of the Cumbre Vieja Volcano, La Palma. Pure. Appl. Geophys. **172**, 3589–3616 (2015)

33. Rampino, M.R., Self, S.: Bottleneck in human evolution and the Toba eruption. Science **262**, 1955 (1993)

34. Lane, C.S., Chorn, B.T., Johnson, T.C.: Ash from the Toba supereruption in Lake Malawi shows no volcanic winter in East Africa at 75 ka. Proc. Natl. Acad. Sci. USA **110**, 8025–8029 (2013)

35. Koulakov, I., et al.: The feeder system of the Toba supervolcano from the slab to the shallow reservoir. Nat. Commun. **7**, 12228 (2016)

36. Kious, W.J., Tilling, R.I., Geological Survey (U.S.): This dynamic earth: the story of plate tectonics. U.S. Geological Survey (1994)

37. Lowenstern, J.B., Smith, R.B., Hill, D.P.: Monitoring super-volcanoes: geophysical and geochemical signals at Yellowstone and other large caldera systems. Philos. Trans. R. Soc. A Math. Phys. Eng. Sci. **364**, 2055–2072 (2006)

38. Archibald, J.D.: What the dinosaur record says about extinction scenarios. Geol. Soc. Am. Spec. Pap. **505**, 213–224 (2014)

39. Kirp, D.L.: Almost home: America's love-hate relationship with community. Princeton University Press (2002)

40. Ritchie, H., Roser, M.: Natural disasters, empirical view. Our World in Data (2018). https://ourworldindata.org/natural-disasters. Last accessed 9 Mar 2019

41. What are the odds of being killed in a tornado?|discover the odds. Discover the Odds. https://discovertheodds.com/what-are-the-odds-of-being-killed-in-a-tornado/. Last accessed 9 Mar 2019

42. Kleinman, Z.: MySpace admits losing 12 years' worth of music uploads—BBC News. BBC News (2019). https://www.bbc.com/news/technology-47610936. Last accessed 7 Apr 2019

43. Transition Town Manifesto. https://www.transitiontowntotnes.org/about/what-is-transition/what-is-resilience/. Last accessed 14 Nov 2016

44. Campbell, C.J., Laherrere, J.F.: The end of cheap oil. Sci. Am. 80–86 (1998)

45. Bardi, U.: The mineral economy: a model for the shape of oil production curves. Energy Policy **33**, 53–61 (2005)

46. Bardi, U.: Peak oil, 20 years later: failed prediction or useful insight? Energy Res. Soc. Sci. **48**, 257–261 (2019)

47. Friedemann, A.J.: When Trucks Stop Running. Energy and the Future of Transportation. Springer, Berlin (2016). https://doi.org/10.1007/978-3-319-26375-5_1

48. Sondland, G.: The fight for EU energy security (2019). https://cassandralegacy.blogspot.com/2019/07/a-rare-glimpse-of-how-elite-think-what.html. Last accessed Aug 27 2019

49. Höök, M., Bardi, U., Feng, L., Pang, X.: Development of oil formation theories and their importance for peak oil. Mar. Pet. Geol. **27**, 1995–2004 (2010)

50. Bardi, U.: peak oil in a fact-free world: the new "oil bonanza" in West Texas. Cassandra's Legacy (2016). https://cassandralegacy.blogspot.com/2016/11/peak-oil-in-fact-free-world-new-oil.html. Last accessed 9 May 2019

51. Arthur, B.: Permian giant oil field would lose $500 billion at today's prices—Art Berman. Art Berman (2016). http://www.artberman.com/permian-giant-oil-field-would-lose-500-billion-at-todays-prices/. Last accessed 9 May 2019

52. Odell, P.R.: Oil and Gas: Crises and Controversies 1961–2000. Multi-Science Pub. Co (2001)

53. Wack, P.: Scenarios: uncharted waters Ahead. *Harv. Bus. Rev.* (1985). https://hbr.org/1985/09/scenarios-uncharted-waters-ahead

54. We are on the cusp of the fastest, deepest, most consequential disruption of transportation in history. RethinkX (2018). https://www.rethinkx.com/headlines . Last accessed 10 Mar 2019

55. Bardi, U.: Cassandra's legacy: crimea: from world war 0 to world war III. Cassandra's Legacy (2017). https://cassandralegacy.blogspot.com/2017/04/crimea-from-world-war-0-to-world-war-iii.html. Last accessed 12 Jan 2019

56. Pinker, S.: The Better Angels of Our Nature. Viking Books (2011)

57. Rummel, R.J.: War & Democide: Never Again. Lumina Press (2004)

58. Pearl, J., Mackenzie, D.: The Book of Why: The New Science of Cause and Effect. Basic Books (2018)

59. Richardson L.F.: Statistics of Deadly Quarrels. The Boxwood Press (1960). https://doi.org/10.1371/journal.pone.0048633

60. Hess, G.D.: An introduction to Lewis Fry Richardson and his mathematical theory of war and peace. Confl. Manag. Peace Sci. **14**, 77–113 (1995)

61. Clauset, A.: Trends and fluctuations in the severity of interstate wars. Science Advances 21 Feb 2018: **4**(2), eaao3580, DOI: https://doi.org/10.1126/sciadv.aao3580

62. Martelloni, G., Di Patti, F., Bardi, U.: Pattern analysis of world conflicts over the past 600 years (2018). arXiv:1812.08071v4

63. Brecke, P.: Conflict (2011). http://brecke.inta.gatech.edu/research/conflict/. Last accessed 17 Nov 2018

64. de Waal, F.B.: Primates–a natural heritage of conflict resolution. Science **289**, 586–590 (2000)

65. Ebert, R.: Night of the living dead movie review (2004). https://www.rogerebert.com/reviews/the-night-of-the-living-dead-1968. Last accessed Aug 27, 2019

66. Rafferty, T.: The state of zombie literature: an autopsy. The New York Times (2011).   https://www.nytimes.com/2011/08/07/books/review/the-state-of-zombie-literature-an-autopsy.html. Last Accessed 27 Aug, 2019

67. Ehrlich, P.R.: The Population Bomb. Sierra Club/Ballantine Books (1968)

68. Darmon, N., Drewnowski, A.: Does social class predict diet quality? Am. J. Clin. Nutr. **87**, 1107–1117 (2008)

69. Stanhope, K.L.: Sugar consumption, metabolic disease and obesity: the state of the controversy. Crit. Rev. Clin. Lab. Sci. **53**, 52–67 (2016)

70. Lee, B.Y., et al.: A systems approach to obesity. Nutr. Rev. **75**, 94–106 (2017)

71. Hasell, J., Roser, M.: Famines. Our World in Data (2018). https://ourworldindata.org/famines. Last accessed 11 Jan 2019

72. Mousseau, F.: Food aid or food sovereignty? (2005). http://www.oaklandinstitute.org/sites/oaklandinstitute.org/files/fasr.pdf. Last accessed 11 July 2016

73. Rummel, R.: Freedom, democide, war. https://www.hawaii.edu/powerkills/. Last accessed 17 Jan 2019

74. Wackernagel, M., et al.: National natural capital accounting with the ecological footprint concept. Ecol. Econ. **29**, 375–390 (1999)

75. Fukuyama, F.: The End of History and the Last Man. Free Press (1992)

76. Orlov, D.: Reinventing Collapse: The Soviet Experience and American Prospects. New Society Publishers (2011)

77. Harper, K.: How climate change and disease helped the fall of Rome—Quartz. Quartz (2017). https://qz.com/1159654/climate-change-helped-destroy-the-roman-empire/. Last accessed 8 June 2019

78. Buntgen, U. et al.: 2500 years of European climate variability and human susceptibility. *Science (80-.)*. **331**, 578–582 (2011)

79. Tainter, J.A.: Complexity, problem solving, and sustainable societies. In: Getting Down to Earth—Practical Applications of Ecological Economics. Island Press (1996)

80. Bardi, U., Falsini, S., Perissi, I.: Toward a general theory of societal collapse: a biophysical examination of tainter's model of the diminishing returns of complexity. Biophys. Econ. Resour. Qual. **4**, 3 (2019)

81. Bardi, U.: The limits to growth in the Soviet Union. Cassandra's Legacy (2015). https://cassandralegacy.blogspot.com/2015/08/the-limits-to-growth-in-soviet-union.html. Last accessed 1 Feb 2019

82. Rindzeviciute, E.: Toward a joint future beyond the iron curtain: the east-west politics of global modelling. In: Andersson, J., Rindzevičiūtė, E. (eds.) The Struggle for the Long Term in Transnational Science and Politics: Forging the Future, pp. 115–143. Routledge (2015)

83. Reynolds, D.B.: Cold War Energy: The Rise and Fall of the Soviet Union. Alaska Chena LLC (2016)

84. Orlov, D.: The Five Stages of Collapse. New Society Publishers (2013)

85. Lippert, B.: Classic ad review: listerine and the halitosis hallelujah|AdAge. Ad Age (2017). https://adage.com/article/classic-ad-review/listerine-halitosis-hallel/310647. Last accessed 5 Apr 2019

86. Anthony Leiserowitz, Edward Maibach, Connie Roser-Renouf, Seth Rosenthal, Matthew Cutler and John Kotcher, Climate change in the american mind, Yale Program on Climate Communication. Report Apr 17, 2018. https://climatecommunication.yale.edu/publications/climate-change-american-mind-march-2018/. Last accessed 27 Aug 2019

87. McPherson, G.: Nature bats last—our days are numbered. Passionately pursue a life of excellence. Nature Bats Last (2019). https://guymcpherson.com/ . Last accessed 6 Apr 2019

88. Goldblatt, C., Watson, A.J.: The runaway greenhouse: implications for future climate change, geoengineering and planetary atmospheres. Philos. Trans. A. Math. Phys. Eng. Sci. **370**, 4197–4216 (2012)

89. Bond, D., Wignall, P.: Large igneous provinces and mass extinctions: an update. Geol. Soc. Am. **505**, 29–55 (2014)

90. Van de Schootbrugge, B., Wignall, P.B.: A tale of two extinctions: converging end-Permian and end-Triassic scenarios. Geol. Mag. **153**, 332–354 (2016)

91. Steffen, W., et al.: Trajectories of the Earth system in the anthropocene. Proc. Natl. Acad. Sci. **115**, 8252–8259 (2018)

92. Ceballos, G., et al.: Accelerated modern human–induced species losses: Entering the sixth mass extinction. Sci. Adv. **1**, e1400253 (2015)

93. Lenton, T.M.: Early warning of climate tipping points. Nat. Clim. Chang. **1**, 201–209 (2011)

4

Strategies for Managing Collapse

Hence to fight and conquer in all your battles is not supreme excellence; supreme excellence consists in breaking the enemy's resistance without fighting.

Sun Tzu, the Art of War

Technological Progress Against Collapse. The Cold Fusion Miracle that Wasn't

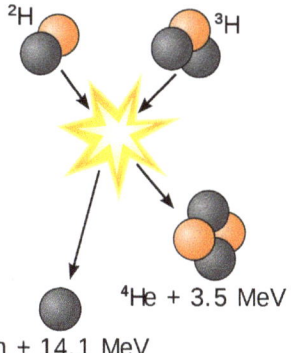

Fig. 4.1 The fusion of a nucleus of deuterium and a nucleus of tritium is believed to be usable as an energy source but it occurs at significant rates only at very high temperatures. In 1989, Martin Fleischmann and his coworker Stanley Pons claimed to have been able to attain the fusion of two deuterium nuclei inside a test tube at near room temperature. It was the dream of "Cold Fusion" that turned out to be just that: a dream (Image from Wikimedia, https://en.wikipedia.org/wiki/Nuclear_fusion#/media/File:Deuterium-tritium_fusion.svg)

© Springer Nature Switzerland AG 2020
U. Bardi, *Before the Collapse*,
https://doi.org/10.1007/978-3-030-29038-2_4

In March 1989, Martin Fleischmann and Stanley Pons, researchers in electrochemistry at the University of Utah in Salt Lake City, published their claim about having attained the room temperature fusion of deuterium nuclei by means of an electrochemical process [1]. It was a new field of nuclear science that they dubbed "cold fusion." If it was true, it was not just the discovery of the century, it was the discovery of the millennium: with their test tubes, Fleischmann and Points had succeeded, it seemed, in tapping the same energy that makes stars burn. It was a discovery that could put to rest all fears of running out of oil at a time when the memory of the great oil crisis of the 1970s was still fresh (Fig. 4.1).

In the months that followed the announcement, almost every scientist in the world who had some background in solid state physics or electrochemistry stopped doing whatever they were doing to examine the new discovery. I was part of that crowd: that year, in July, I traveled to California to spend the summer to work at the Lawrence Berkeley Laboratory. There, they had one of the best surface science and electrochemistry labs in the world and if anyone was able to confirm the claims of cold fusion, they were the right ones.

When I arrived in Berkeley, I expected to find my colleagues excited by the new discovery and maybe working on it. But I found that they had already passed that stage and now they were disappointed. They had tried to replicate the cold fusion experiments without getting any results. They had concluded that the whole story was a mistake or, worse, a scam. So, I spent that summer in Berkeley working on subjects not related to cold fusion, but I had not given up: the fascination of the idea of being able to replicate a star in a test tube was too strong. So, back to Italy, in September, I thought I could do some experiments myself using a different setup than the one that my colleagues in Berkeley had experimented with. Maybe, in that way, I could see something that they had missed.

Let me not bother you with the details of what I did, here, you can find a little more in a blog post of mine [2]. Let me just tell you that I spent a few months working alone in my lab, feeling a little like Dr. Zarkov, the character of the *Flash Gordon* comics, who builds a spaceship in his basement.

But, in my case, no spaceship emerged out of the lab. I soon discovered that if there was such a thing as "cold fusion" it was a very weak effect, if it was there at all. For sure it was nothing like the strong effect that Fleischmann and Pons had claimed when they spoke of the "ignition" of the deuterium they were using in their experiments. No matter what I tried to do, I could not see anything like that with my setup.

I did not give up immediately, there was a certain "Elvis sighting" atmosphere about cold fusion at that time. It was not unlike the many claims of having seen Elvis Presley alive in the 1980s, after he died in 1977. Claims of experimental evidence of cold fusion were popping up everywhere and that made me think that maybe I was a bad experimenter, that I was making some mistake. The Elvis sighting effect can be strong: you tend to see what other people claim to have seen. Several times I thought I had seen a signal that showed that, yes, a nuclear reaction was taking place in the steel vessel I was using for the test. It seemed that, really, the energy that powers stars had appeared in my lab. But when I redid the experiment, the signal was gone. I was chasing a ghost and, by Christmas of 1989, I gave up.

Rethinking about that old story, I think I was lucky that I lost just a few months of work. Others would spend years, stake their reputation on some uncertain results, and retire decades later still claiming that the elusive room temperature fusion was just one more experiment away. One of the characteristics of "pathological science", indeed, is that the signal is always weak, at the edge of the sensitivity of the instrumentation. But only pathologically optimistic scientists could see that signal and, gradually, cold fusion slipped away from science to settle into something performed by colorful figures of pseudo-scientists or mad solitary geniuses touting weird machines and claiming that they are going to revolutionize the world. But that is always for next year, or for as soon as the new machine or the new test is ready. Changing the name of a discredited field did not help: turning "cold fusion" into the more hi-sounding "LENR" (low energy nuclear reactions) did not change the fact that nuclear fusion is not and cannot be "low energy." Call it the way you like, cold fusion or LENR, it turned out to be full of sound and fury, signifying nothing.

Gradually, interest in the idea faded but, even today, people are still fascinated with the idea of reproducing a star in a test tube. So, 30 years after the first claims by Fleischmann and Pons, the communication giant Google engaged some researchers in a program aimed at trying again to find signs of nuclear fusion at near room temperature [3]. Unsurprisingly, they found nothing: they just repeated experiments that had already been done, confirming that there is no such thing as "cold fusion" (or LENR). They might as well have sent their researchers to search for the lost ark of the covenant.

This enthusiasm for something that does not exist was always fueled not so much because it was a new physical phenomenon: nuclear fusion had been known for at least half a century. Cold fusion was always presented as something that would fulfill the prophecy of the 1950s that nuclear technologies would bring to us energy "too cheap to meter." It was a prophecy borne out of the incredible achievements of the 1940s and 1950s, when it really seemed that nuclear energy was a Pandora's box that would bring to us perpetual abundance. No one who has watched Walt Disney's movie *Our Friend, the Atom* (1957) as a teenager can forget the atmosphere of expectation of great things to come of those years.

But reality was, as usual, around the corner and the promise of nuclear fission turned out to be much less exciting than it had seemed to be at the beginning. Apart from accidents, the problem of proliferation, the difficulties of controlling the technology, it was soon discovered that the mineral reserves of uranium were far from sufficient for the kind of limitless prosperity that had been imagined at the beginning. If we wanted enough fuel for the kind of abundance envisioned in the 1950s, we would have had to engage in the dirty and dangerous business of "breeding" nuclear fuels in the form of plutonium to make up for the scant uranium resources. But the idea was soon abandoned: too complex, expensive, and risky in political terms. Nobody wanted plutonium to become commonplace all over the world when it could be used to make nuclear warheads or, more simply, turned into a deadly poison. That left nuclear fusion as the workhorse of nuclear hopes: the energy that powers stars. It seemed obvious that, if we could have it here, on Earth, all problems with energy would fade away forever.

Alas, controlled nuclear fusion turned out to be an elusive dream. It is not impossible to attain it on our planet: it can be done inside nuclear warheads, but that is not the kind of technology you can use to power the electric grid. What people were dreaming about was the concept of "controlled" nuclear fusion, the same kind of taming of the enormous nuclear energies that had been obtained with nuclear fission. In the 1950s, it seemed to be just the next step in an unstoppable progression of better technologies, but things turned out to be more difficult than imagined. Decades of work and untold billions of dollars were spent to build larger and larger "Tokamak" machines supposed to be able to reach temperatures so high that "hot" nuclear fusion would take place at a sufficiently fast rate for useful energy to be produced. So far, the only result obtained was to show that a bigger machine was needed. The latest incarnation of this "big is beautiful" approach is the ITER machine, being built in Southern France. It is so big that 35 nations had to pool their resources in order to make the project possible. Construction was started in 2007 and the machine is scheduled to start working as a fusion reactor by 2035 [4]. That doesn't mean that ITER will produce useful energy—a new and even bigger machine will be needed for that—if it will ever work. It is even more uncertain whether it will make economic sense to use it. At this rate, our civilization may go through a couple of Seneca Cliffs before we find a way to make this kind of machines useful for something.

Other approaches to fusion not based on tokamaks turned out to lead to dead ends, too. It is often possible to create devices that can produce nuclear fusion, the problem is to turn them into useful energy sources. There may be a fundamental problem here: despite all the hype, it might be that nuclear fusion is just not such a great idea for what we need. The power density of the Sun is ridiculously low: less than 300 Watts per cubic meter [5]. The engine of a small car may have a power density thousands of times larger! Nature, it seems, doesn't like to keep very high power densities for long times and stars are spectacular machines but not very efficient ones. So, the dream of cheap and abundant energy from nuclear reactions may always remain a dream, at least on our planet.

But let us crank up the dreaming machine into motion and start speculating a little. What if we could really develop a miraculous technology that would give us nearly free, non-polluting, and abundant energy? Would that help us avoid the impending Seneca cliff of our civilization?

First of all, with cheap and abundant energy, the depletion of mineral resources would not be a problem. We would not need anymore to mine from depleting ores, we could just mine the crust for whatever element we need. It would be the concept of the "universal mining machine" [6], a mechanism that eats rocks and spits out their contents nicely arranged in boxes of pure elements. A machine like that is physically possible but, today, it would make no sense because of horrendous costs in terms of the energy it would need. But what if we could increase the global energy supply by a factor, say, one hundred or one thousand? Then, we could really mine the Earth's crust to obtain all the chemical elements we need. Of course, these machines would also produce a gigantic amount of pollution but they could be sent to the Moon or to the asteroids and the pollution would remain there while the precious materials mined could be shipped to Earth. Or, with abundant energy, we could ship pollution to space.

Then, how about the problem of human overpopulation? Cheap and abundant energy could solve that problem, too. We could use artificial light to power photosynthesis on a truly gigantic scale. There is a wonderful science fiction novel by Robert Hanson Heinlein, *The Moon is a Harsh Mistress* (1965) describing a future in which the Moon has become a granary for an ever-expanding Earth population, with the grain shipped to Earth by means of an "electromagnetic catapult." If something like that were possible, we could turn the Earth into a planet similar to Trantor, the galactic capital described in Isaac Asimov *Galactic Cycle*: a completely urbanized planet formed of a single, giant city, covering the whole landmass. Then we could have hundreds of billions of people on Earth and, probably, no other species of body mass larger than a few kgs except, perhaps, for cows. Maybe cows could be raised on the Moon, too.

If we had really large amounts of cheap energy, we could ship people to space and have them live inside giant artificial habitats orbiting around the Earth, a daring scheme proposed in 1974 by Gerard O'Neill [7], in part as a response to the scenarios of collapse proposed in the first edition of *The Limits to Growth,* in 1972. O'Neill's concept was based on immense pressurized habitats that would be placed at the L4 and L5 Lagrange points, where the interplay of the gravitational fields of the Moon and the Earth, and the Sun generates a minimum in the gravitational potential. At these points, an object can remain in a stable position in principle forever. Some dreams of space colonization turned out to be even grander. In 1960, Freeman Dyson [8] proposed that the whole Solar system could be turned into an immense sphere surrounding the Sun, built using matter obtained from dismantling the planets. If such a feat were possible, it would increase the human habitat by an enormous factor in comparison to occupying the surface of just one planet. Some other studies even considered the possibility of colonizing the whole galaxy. Although the speed of light is an absolute limit that, as far as we know, cannot be overcome, even at relatively slow speeds, an intelligent species could colonize the galaxy in times of the order of a million years [9].

The concept of unlimited energy available can be modeled and it was done for the first time in the 1972 study "*The Limits to Growth,*" [10]. The model used did not consider energy as a disaggregated parameter but it could be indirectly modeled by removing the limits to the flux of natural resources into the economy. A simulation along these lines was performed already in the first *Limits* study, in 1972, and it was confirmed in the later versions: infinite energy available postpones collapse but generates it anyway as the result of a combination of overpopulation, depletion of agricultural soil, and pollution. But, if these limits are removed, too, assuming an expansion into space, then we have a scenario that the authors of the study termed IFI-IFO (infinite in, infinite out). And, as you would expect, the result is that the economy and the human population keeps growing forever or, at least, for as long as you care to run the model into the future. Yes, but also Santa Claus could solve a lot of problems if he existed.

So, let's go back to the real world and examine what we could reasonably do in terms of technological progress to avoid the Seneca Cliff for our civilization or, at least, mitigate its damage. Of course, we must first ask ourselves what we mean as progress. Spaceships? Smartphones? Laser beams? Boner pills? All this and more, but what is it that links together all those things? How can we define progress? And how can we measure it when we are not sure how to define it? One thing we can say about it is that it is a relatively new idea: the ancient Romans or the people of the Middle Ages would see no difference in their way of living compared with that of their parents or grandparents, and not even for people living centuries before. They would have been baffled by the concept that, somehow, tinkering with mechanical things would change their lives and make the world better. It was only during the 18th century that Edward Gibbon noted the trend of technological progress perhaps for the first time in history his *Decline and fall of the Roman Empire* [1788] when he wrote that, "*The ancients were destitute of many of the conveniences of life which have been invented or improved by the progress of industry.*" In time, the concept of progress became commonplace and the enthusiasm for progress probably spiked up to the highest level during the mid-20th century, when the "Atomic Age" was in full swing and people expected friendly home robots, flying cars, and weekends on the Moon for the whole family. The mid-20th century was also the time when the first attempts at quantifying progress were performed.

The merit of having been the first to try to quantify progress goes perhaps to Robert Anson Heinlein (1907–1988) mainly known as a science fiction writer. In his 1952 article titled *Pandora's box* (originally published with the title *Where To?* [11]) he proposed that technological progress had been growing exponentially up to then and would continue to grow exponentially in the future, bringing unimaginable wonders to humankind. It was a bold attempt to understand a difficult concept, but also flawed in many ways. Heinlein did not even attempt to define or quantify his concept of "technological progress," he just drew by hand a growing curve on a Cartesian graph. Then, his detailed predictions turned out to be nearly all wrong. He spoke of anti-gravity, space flight for the masses, life extension over 100 years for humans, and many other wonders that never materialized. On the contrary, he failed to imagine such things as the Internet, cell phones, personal computers and most of what we consider today as the tangible manifestations of progress.

But the idea that technology grows exponentially seemed to be mature in the 1950s and it appeared in a different form when, in 1956, the economist Robert Solow published the results of a study that is often considered the basis of the understanding of technological progress in economics [12]. Solow could fit his data assuming the presence of a factor, that he called "*A(t)*," that grew exponentially with time. This entity came to be known as "Solow's residual" or "Total Factor Productivity" (TFP) and it is commonly understood as a quantitative measurement of technological progress. According to Solow, it grows exponentially with time at a rate of 1%-2% per year. If this factor could keep growing forever, it would easily compensate for such factors as the decline of the availability of natural resources, as argued, for instance, by William Nordhaus in 1992 [13]. Just 1%–2% per year? That does not seem to be so difficult. If we could keep that rate of growth of progress, the A (t) factor would get rid of all cliffs and keep the economy growing forever or, at least, for a very, very long time. That is surely a comforting idea, and it is by now rather well entrenched in economics and with policymakers. So much that when a problem appears, the knee-jerk reaction of many politicians is "we must finance more research."

But is it true that progress grows exponentially with time? And what is exactly this "Solow residual?" How can we be sure that it will keep growing exponentially, assuming that is what it has been doing up to now? And can we put our trust in a parameter that cannot be measured but can only be inferred on the basis of a highly simplified model. The residual identified by Solow may actually exist, but it may be related to factors other than technological progress. It may simply be proportional to the supply of energy to the system, as proposed, among others, by Robert Ayres [14]. So, the incorporeal TFP factor may really be something much more concrete than what it was thought to be. Indeed, the conventional understanding of the TFP was criticized by Herman Daly in his *Steady state Economics* (1977) [15] where we can read in chapter 5 that:

> The idea that technology accounts for half or more of the observed increase in output in recent times is a finding about which econometricians themselves disagree. For example, D. W. Jorgenson and Z. Grilliches found that "if real product and real factor input are accurately accounted for, the observed growth in total factor productivity is negligible" (1967). In other words, the increment in real output from 1945 to 1965 is almost totally explained (96.7 percent) by increments in real inputs, with very little residual (3.3 percent) left to impute to technical change. Such findings cast doubt on the notion that technology, unaided by increased resource flows, can give us enormous increases in output. In fact, the law of conservation of matter and energy by itself should make us skeptical of the claim that real output can increase continuously with no increase in real inputs.

A further perplexity on the role of the TFP residual derives from the fact that it may be the only entity in economics that is supposed to keep growing forever. That is curious, to say the least, considering the established concept of "diminishing returns" in economic sciences. Why should technological progress be exempt from this very general law? This point was examined already in the 1970s by Giarini and Laubergé [16] and more recently by Tainter [17]. From these studies, it seems clear that the growth rate of technological progress is slowing down in our times. It is not growing exponentially anymore, assuming that it did in the past.

There are plenty of technological areas progressing very slowly if they are progressing at all. Just think of how the human average life expectancy is not significantly increasing any more after the spectacular rise observed up to a few decades ago. Even highly touted cases, such as "Moore's law" in electronics, are showing signs of fatigue. Moore's law indicated the number of elements placed on a computing chip should double every two years, approximately. But it has been clearly slowing down—perhaps just disappearing—during the past few years [18]. The mysterious technological force that is said to push the economy onward may be made of such stuff as cold fusion is made of: dreams and bad measurements.

That does not mean that technological progress does not exist, but it means that we need to look at it as something real, something that works, something other than uncertain parameters of uncertain models. What kind of technology do we need to avoid the Seneca Cliff we are facing?

Nowadays, much research is about solutions that would worsen the problem. Think of biofuels: they are another knee-jerk solution to depletion problems. "Are we running out of oil?" So, what's the problem? We'll use biofuels! But that makes no sense if you think of it quantitatively. Photosynthesis, the process plants use to create organic molecules out of sunlight and atmospheric carbon dioxide, is not very efficient, around 1% on the average, probably less than that for crops. So, it is easy to calculate that if we were to use agriculture to produce the fuel needed for the gigantic fleet of fossil fuel-powered vehicles of today, we would use most of the available agricultural land [19]. And, surely, the idea of starving people in order to feed cars does not seem to be very smart. So far, the effort on biofuel cultivation has resulted mainly in the wholesale destruction of many primeval forests to cultivate palm oil and, as a consequence, to the near extinction of orangutans. All that just for the production of little more than 2% of the total diesel fuel produced in the world [20]. Maybe you do not care about the Seneca collapse of orangutans, but for sure it will not save us from our own collapse. So, is it worth it?

Similar considerations can be made for the many efforts to develop technologies making us more energy efficient. That is surely a worthy task in many respects. It is a good thing to insulate our homes, use more efficient cars, LED lights, public transportation, organic food, and things like that. But would it save us from depletion and climate collapse? Unfortunately, in many cases all these efficiency-related ideas amount to little more than greenwashing. Not that they are bad ideas, but their economic return is slow: it takes several years to recover the investment in, say, insulating one's house. And we are running out of time with mineral depletion and climate change.

Then, there is a perverse effect associated with technologies that improve efficiency. You probably heard of the "Jevons Paradox," described for the first time in Jevons' 1865 book *The Coal Question* [21]. The gist of Jevons' idea was that improvements in efficiency do not lead to a reduction in the amount of energy used, something that he could demonstrate by means of data on the use of coal-powered steam engines in England during the 19th century. It is not obvious that the "paradox" holds exactly in its original form in modern times, but studies tend to support this idea [22] under such names as "rebound," "backfire," and "Khazzoom-Brookes Postulate." Indeed, the idea makes a lot of sense: it is not at all a paradox. Imagine that you insulated your home: it means you save money in heating costs and what will you do with that money? Maybe you'll make a donation to the WWF to save the tortoises of the island of Pago-Pago but, more likely, you will take a vacation to Hawai'i using at least the same amount of fossil resources and creating the same amount of pollution that you would have created by means of your heating system before insulating your home.

This discussion may sound pessimistic but we do not have to be discouraged, we only need to be more creative. If technology cannot produce miracles, it is also true that maybe we do not need them. We saw that complex systems are entropy-producing machines that feed on energy potentials. So, if we want the complex system we call "civilization" to keep going in some form or another, we need to provide food for it: an amount of energy comparable to the one produced today mainly by means of fossil fuels. It is not impossible. The paper that myself, Sgouris Sgouridis, and Denes Csala published in 2016 with the title The Sower's Way [23] shows that the renewable technologies we have today, mainly wind and photovoltaics, are good enough to replace the energy flow we obtain today from the dwindling fossil fuel resources, without causing greenhouse emissions. We found also that it would be possible to use the remaining fossil fuels to jump start a renewable-based infrastructure that, subsequently, would not need fossil fuels anymore. In other words, we would use fossil fuels in the same way as our farmer ancestors used corn saved from the previous harvest for the new one. A nice idea with one glitch: it will be very expensive, although not impossible. The data also show that, if we want this transition, we have to start paying for it right now. We need to increase by about a factor of 50 the amount of energy invested in creating a new energy infrastructure. That is unlikely to happen considering that in the present debate the opinion leaders have not yet realized the true potential of renewable energy. Apparently, we are not as wise as our ancestors and we believe that the good thing to do is to eat our seed corn. As long as we keep this attitude, no technological progress will save us from the coming Seneca Cliff.

To conclude this chapter, let me note that there exists another view of technological progress, grander and more ambitious than the one that derives from the smooth curves of economics models. As an example of this view, we can cite Kevin Kelly's book *Out of Control* [Kelly 1994] where we find a description of progress that was produced as a direct criticism of the *Limits to Growth* study. We read at p 575 that:

> Direct feedback models such as Limits to Growth can achieve stabilization, one attribute of living systems, but they cannot learn, grow or diversify—three essential complexities for a model of changing culture or life. Without these abilities, a world model will fall far behind the moving reality. A learning-less model can be used to anticipate the near future where co-evolutionary change is minimal; but to predict an evolutionary system—if it can ever be predicted in pockets—will require the exquisite complexity of a simulated artificial evolutionary model.

And:

> *The Limits of Growth* cannot mimic the emergence of the industrial revolution from the agrarian age. "Nor," admits Meadows, "can it take the world from the industrial revolution to whatever follows next beyond that.

In this view, progress is something that moves in leaps and bound, actually in "quantum leaps," and as it grows it spikes up changing everything radically and forever. From a human viewpoint, at some moment, progress it will appear to, literally, shoot out to infinity. In some interpretations, this phenomenon will lead humankind to transcend into a nearly godlike, "transhuman" status, an idea that may have been expressed for the first time in its modern form with Robert Ettinger's book "*man into Superman,* originally published in 1972 [24]. The most recent proposer of the concept of technological singularity is probably Ray Kurzweil, who has published several books on the subject. Among these *The Singularity is near* [25]. These concepts are fascinating but, at present, they remain in the realms of possibilities for the future. If humankind goes through a technological singularity, then we cannot know where it will go, and not even if it will continue existing afterward.

Even without these extreme possibilities, it is clear that technology in its expression of Artificial Intelligence (AI) is taking us somewhere, and that somewhere may not be exactly where we want to go. The Web is more and more invading our minds, changing us, rather than changing our environment. Instead of finding the magic energy trick to have abundant energy, it may lead us not to need it. But will it? Let me cite from a recent article by George Dyson on *Edge* [26]

> Most of us, most of the time, are following instructions delivered to us by computers rather than the other way around. The digital revolution has come full circle and the next revolution, an analog revolution, has begun. None dare speak its name.
>
> The genius — sometimes deliberate, sometimes accidental— of the enterprises now on such a steep ascent is that they have found their way through the looking-glass and emerged as something else. Their models are no longer models. The search engine is no longer a model of human knowledge, it *is* human knowledge. What began as a mapping of human meaning now defines human meaning, and has begun to control, rather than simply catalog or index, human thought. No one is at the controls. If enough drivers subscribe to a real-time map, traffic is controlled, with no central model except the traffic itself. The successful social network is no longer a model of the social graph, it is the social graph.
>
> We imagine that individuals, or individual algorithms, are still behind the curtain somewhere, in control. We are fooling ourselves. The new gatekeepers, by controlling the flow of information, rule a growing sector of the world.

What's going to happen to us? Will it alter the way our brains are built, with the ingrained desire to have more? Will it lead us to learn to live with the limits we have? Whatever happens, the future is never like the past: if the next Seneca Cliff will be in real space or virtual space, we cannot say.

The Evil Side of Collapse: The Iago Strategy

Fig. 4.2   The character of "The Joker" in at the 2015 art exhibition at the Barcelona International Comics convention, complete with the Satanic laughter pertaining to truly evil characters (Picture by Ivan Bea, https://en.wikipedia.org/wiki/Joker_(character) #/media/File:Joker_expo.jpg)

With Iago, in *Othello,* William Shakespeare created perhaps the best evil character in the history of literature. The drama is all based on the subtle plotting of Iago to get revenge on his master, Othello, by having him suspect his wife, Desdemona, of betraying him. In the story, neither Othello nor Desdemona are described as especially dumb people, but they are overwhelmed by the superior cunning abilities of Iago who exploits every detail, every chance, every event, to fan Othello's suspicions until, eventually, Othello is led to killing his wife and then to kill himself.

In modern times, it seems that the subtle and sophisticated evil characters of past literature, such as Iago, have been replaced by ugly monsters endowed with little more than a Satanic smile and the kind of laughter that goes like "*Bwa-ha-ha-ha*" in comics. But if evil characters have existed in fiction since the time of the Sumerian priestess Enheduanna, it is because they are the mirror of something real. In your everyday life, you will rarely see the equivalent of "The Joker," the arch-villain of the Batman universe, but you do see equivalents of Iago in terms of people managing the twists and the traps of what we call "office politics." Some people seem to show an uncanny skill in maneuvering things in such a way to damage other people. They can destroy themselves as well! I don't know about your experience, but I saw that happening more than once in my career. And, of course, evil is a common occurrence in politics, where people in positions of power can do a lot of damage to all of us.

Iago is truly the embodiment of the concept of evil in the sense attributed to Satan himself described as "The Master of Lies." How does he attain this proficiency of arch-villain? I would say that Iago masters the science of complex systems. His actions follow the basic tenets of Griffith's theory of fracture: he is engaged in creating small cracks in the network of the social relations among the characters surrounding him, making the fissures grow by exploiting the internal strains of the connections. The cracks grow until they coalesce into a single one in the relation between Othello and Desdemona. The crack grows longer than the Griffith length, and it makes the system go critical and pass through a tipping point: tragedy ensues, as we know. We could call this technique of destroying a complex system "The Iago Strategy."

The idea of using collapse to get rid of your competitors and enemies goes beyond individual actions, and may become a business or a political strategy. Especially in politics, calumny is a well known and honed strategy, sometimes going under the name of "muckraking" when it is done by journalists. In some cases, calumny is part of an election strategy: an example is how Lyndon Johnson damaged his opponent, Barry Goldwater, in the presidential elections of 1964 by accusing him of planning a nuclear war. On a larger scale, the method is part of the concept of "Yellow Journalism," a technique that combines exaggerations, wild claims, and unsupported accusations aimed at specific persons. It became popular in the US starting with the late 19th century, and it is still very popular today. We just need to remember the case of Dominique Strauss-Kahn, French manager and politician, who was accused in 2011 of having sexually attacked a hotel maid in New York. The story was, and continues to be, highly controversial but surely it thwarted his ambitions to compete for the presidency of France.

The idea of causing an opponent to collapse may not refer just to political struggles. As Von Clausewitz said, war is nothing more than the continuation of politics by other means and the capability of causing the collapse of the enemy has obvious military implications. Warfare is, after all, a struggle that involves complex systems: armies fight and maneuver against each other, entire countries support them, the battle goes on and it ends when one of the two sides collapses as the result of accumulated strain.

The most brutal and expensive way to get rid of an enemy is simply to destroy it. But, already in ancient times, Sun Tzu noted how "all warfare is based on deception." That seems to imply that the best way to win a war would be to exploit the internal strains of the enemy's networked structure, and this needs to be done in a covert manner. Then, the enemy will defeat itself and, citing again from Sun Tzu, "the supreme art of war is to subdue the enemy without fighting."

It must be said that, in modern times, these ideas do not seem to be very popular with the military or with politicians. Maybe a wave of barbarism is pervading the world, but the Second World War was the last major war to be formally declared by the governments engaged in it. Afterward, only a few local wars were actually declared despite many having been fought. Nowadays, the war goes on until the losing side is utterly destroyed and its leaders captured and often executed.

Wars may become more cruel and ruthless than they used to be also in another factor: the involvement of civilians. Of course, exterminating civilians is an ancient tradition but, in our times, it is supposed to be illegal and those who directly target civilians risk being hanged when the war is over (of course, only if they are on the losing side). In practice, the idea of civilians as a legitimate war target is deeply entrenched in the current military thought. It seems that it was explicitly proposed for the first time in modern times by Giulio Douhet, Italian officer and the author of "The Command of the Air" (*Il dominio dell'aria*) (1921). Douhet's ideas seem to be taken from an evil character of a comic book, a sort of early "Joker," even though we have no record that Dohuet would intersperse bouts of Satanic laughter within his utterances on strategy. But the concept he proposed was truly evil: abandon all conventional warfare intended as a struggle of armed forces and concentrate instead on aerial bombing to kill civilians. They will have to surrender, else they will be exterminated (Fig. 4.3).

Fig. 4.3 An American B17 Bomber in action over Germany in 1943 (Image from National archives. https://www.archives.gov/files/research/military/ww2/photos/images/ww2-73.jpg)

The idea of killing everyone on the other side is at the basis of the deployment of the various mass murder weapons that were accumulated and sometimes used along the 20th century. Still today, the USA and Russia have considerable overkill capabilities against each other and against the whole humankind in terms of the number of nuclear weapons they stockpile. Other countries may not be able to exterminate humankind by using the nuclear weapons they possess, but they seem to be doing their best efforts in that direction.

In addition to nuclear weapons, there are interesting (in a certain sense) possibilities in terms of mass extermination by means of chemical and bacteriological weapons, although neither seem to have been experimented on a truly large scale, so far. The same is true for the latest generation of hi-tech weapons: aerial drones which might also be used for purposes of extermination. At present, they seem to be only used for "targeted killing" directed against a relatively small number of targets. The latest available data speak of some 10,000 victims of drone strikes carried out by US forces from 2004 to date [27]. We have no idea of how reliable this estimate can be. If it is, this is a relatively small number of casualties, but surely drone warfare could be stepped up and these weapons turned into proper mass murdering tools. The concept of killer microdrones has been described in the 2017 "slaughterbots" movie by the Future of Life institute and Stuart Russell [28]. It is based on the idea of small drones carrying a small explosive charge, sufficient to kill a person, and with facial recognition technologies able to identify specific persons or generic people who wear a certain uniform or have some ethnic facial traits. If that is not evil, I do not know what is. Maybe the makers of this weapon could improve it by adding the capability for the drone to emit a Satanic laughter that goes *Bwa-ha-ha-ha* just before it kills its target by exploding near his or her forehead. Fortunately, it seems that this technology is not available, yet, but there is no reason why it could not be developed in the future.

Mass extermination is surely a way to push an enemy down a steep Seneca Cliff, but it seems to be a little drastic as a method. Besides, it has a big problem that, curiously, Douhet and his followers completely forgot to take into account. If you have an inexpensive and effective technology to kill *them*, chances are that *they* will have it, too, to be used against you. And that makes things a little problematic with the risk of symmetric reciprocal extermination, as nearly happened in Europe during WW2 with aerial bombing in which the Allies and the Axis forces engaged. It is strange that this point does not appear clear either to the public or to policymakers. For instance, a recent survey carried out by the *Bulletin of The Atomic Scientists* [29] finds strong support with the American people for a preventive nuclear attack against Korea that would kill one million people, there. Apparently, many people love the idea of pushing others down what could be the steepest Seneca cliff of all, nuclear extermination, without thinking too much about what the targeted nation could do in terms of retaliation. But killing people on both sides until nobody is left alive looks a little dumb as a military strategy, to say the least. Can't we think of something smarter?

If war is a struggle involving the stability of complex systems, a smart strategy would consist in exploiting the networked structure of the enemy society to cause it to collapse: it is the system science view. An army, or any fighting organization, is a network and in all networks nodes must communicate with each other. So, every army is sensible to collapse caused by a loss of communication and, in particular, to the feedback effect that takes place when the nodes communicate the wrong information to each other, For instance, if a soldier starts running away from the battlefield, soldiers nearby receive the communication that things are not going well and they may start running away, too. Enhancing feedbacks take over and the whole army melts away: it is the nightmare of all generals, ancient and modern.

Avoiding this occurrence is the reason why modern armies are pyramidal networks where each node communicates almost exclusively with the upper and the lower layer. Soldiers do not give orders to each other, they receive them from their officers who in turn receive orders from higher level officers and the whole army depends on a central command. This kind of structure avoids the melting catastrophe but makes the army sensitive to "decapitation strike". If all communication must pass through a single node of the network, then removing this node is a way to generate a Seneca Collapse.

The problem with the idea of destroying a military structure by decapitation is two-fold: the first is that this vulnerability is well known and strategies are normally implemented to ensure that leaders are difficult to kill. For instance, in the United States, the president has a bunker under the White House that's supposed to be used as a secure shelter and communications center in case of an emergency. In case of a major war and of threats against the US territory, the president is expected to be flying in a "doomsday plane," a plane with the sole purpose ofing keep the president in the air, where he is presumably difficult to locate and hence safe.

A different approach to counter the risk of a decapitation strike is to abandon the typically rigid structure of armies and adopt a flexible one with small units able to continue fighting even if they lose contact with their command center. It is a way of fighting that was pioneered by Edwin Rommel on the Italian front during the First World War. A recent example of resilience in an armed conflict is the 2006 confrontation between Israel and Hezbollah in Lebanon, where Hezbollah successfully applied this strategy.

The concept of inducing a collapse in the enemy army is a way to improve the effectiveness of warfare while at the same time reducing the cost and violence of a conflict, but it remains embedded in the conventional views of wars fought by armies. Nowadays, the very idea of conventional armies may be obsolete. War is becoming more and more embedded in the structure of society, taking different shapes under the general concept of "hybrid war." Modern armies are part of a network that includes the economic, social, political, and religious structure of a whole country. Attacking or weakening this larger network may lead it to collapse and, even though the army may maintain its fighting capabilities, it becomes useless without a country to support it.

It is an idea that runs along the lines of the extermination proposal put forward by Douhet, but it is more sophisticated: a hybrid war is not about exterminating civilians, at least not directly. It is about weakening the economic and social structure of an enemy country, if possible causing its collapse so that it cannot support a war effort anymore. A good example is the fall of the Soviet Union in 1991. The Red Army was not defeated, not even attacked, and at the moment of the fall it maintained most of its fighting capability. But there was no government anymore able to pay the salaries of soldiers and officers. So, the army went through a Seneca collapse and dissolved.

Economic warfare is a common component of hybrid warfare. It may take different shapes: in its most brutal form it simply consists in starving the enemy population, to death if necessary. There are many examples of this strategy being applied in ancient times. We have a poignant example in the description of the siege of Jerusalem in 70 CE by Flavius Josephus where he tells us of such graphic details as mothers eating their children. In modern times, we may remember how, in 2018, the US secretary of State, Rex Tillerson, declared that the economic sanctions enacted against North Korea imposed from 2006 are effective because of the evidence of deaths caused by starvation in the country [30].

A specific variant of economic warfare is "energy warfare," consisting in starving an enemy country not of food but of energy. It may have been tried for the first time by the Allies with their attack on German dams carried out in 1943 in the "Operation Chastise" carried out using a purpose-built "bouncing bomb" designed to skim over the surface of the German hydroelectric basins before detonating against the dam wall. The attack was successful in the sense that it caused considerable damage to German dams, but it had little long term effects and it cost to the allies 40% of the attacking aircraft.

Another case was the Israeli air strike carried out on 7 June 1981 which destroyed an Iraqi nuclear reactor southeast of Baghdad—the plant was still under construction and held no nuclear material. Later on, the Iraqis targeted an Iranian nuclear reactor in Bushehr in 1987. Neither strike had a significant military effect. Then, the 1999 NATO bombing campaign against Yugoslavia saw attacks specifically directed against power plants. During the early phases of the campaign, NATO planes used a special "soft bomb" or "graphite bomb," specifically created to emit a cloud of graphite to short-circuit the connections of power plants [31]. The Western press reported that these bombs disabled about 70% of the Serbian electric grid. The Serbians admitted that they experienced blackouts, but also claimed that they were able to restore power in a short time and that the effect of the attacks was negligible. We do not seem to have a reliable assessment of the actual results of the attacks and, in any case, after that first attack, NATO did not use any more graphite bombs, preferring to use conventional weapons directed against power plants and transformer stations. None of these attacks succeeded in forcing Serbia to surrender and so far, the idea of targeting the energy network of a whole country has never been very effective. But, if it were to succeed on a large scale, the consequences of leaving a whole country without power for a long time would be so devastating as to be nearly inconceivable, a Seneca Collapse that nobody would ever want to see.

Overall, the simplest way to cause economic damage to an enemy population is by means of economic sanctions. That may be a very powerful weapon and it can starve whole countries although, in modern times, it seems that sanctions are rarely carried out to their extreme consequences. For instance, the economic embargo against Iraq after the first gulf war in 1991 was relaxed to allow Iraq to export oil in order to import food and avoid mass starvation of its population.

In general, the idea at the basis of all hybrid war methods is that the targeted civilian population should not be exterminated, but rather become discouraged and cease to support the war effort. In history, that turned out to be difficult and often counterproductive. Starved or bombed people will normally direct their hate toward those who are starving or bombing them, not necessarily against their government, no matter how oppressive and dictatorial it is. If you want an example of how economic sanctions may misfire, consider the case of the international sanctions against Italy imposed by the League of the Nations in 1935–1936 [32], after that Italy had invaded Ethiopia. The sanctions generated strong nationalistic feelings in the country and reinforced the grip of the Fascist Party on the government. Later on, when Britain enforced a coal embargo against Italy, the result was that Germany became the main supplier of coal to Italy and that led Italy to join Germany during WW2 [33]. Embargoes seem to normally achieve exactly the opposite effect of what they are said to be enacted for. Or, possibly, this is exactly what they are enacted for: to force a country to go to war even in unfavorable conditions.

So, it seems that if we want to cause the collapse of an enemy without the need of conventional warfare, we need something subtler and more effective than bombs or economic sanctions: we need to convince the population of the target country that their enemy is their own government. This is the basis of the subset of hybrid warfare known as "psyops" (psychological operations). It is a way of waging war that mainly relies on propaganda, but with a few extra twists. Normally, propaganda takes a reactive approach, trying to influence people's perception of reality by means of three cardinal techniques: obfuscation (denying or hiding information), saturation (distracting the targets by means of irrelevant information) and spin (presenting information in a form favorable to a certain interpretation) [34]. Acting along these lines, propaganda is a consensus-building technology used mainly as a tool for reinforcing national cohesion. That is often obtained by developing hate against some political, ethnic, or religious enemy.

Psyops use some of the typical techniques of propaganda, but they are more aggressive and tend to be pro-active in stimulating some kind of action. They are probably best described in terms of a quote attributed to an "aide of the Bush administration" at the time of the 2003 invasion of Iraq in an article by Ron Suskind in *The New York Times*, in 2004 [35]. The quote is often attributed to Karl Rove, although Rove himself denied being the author. It is worth reporting it in full:

> The aide said that guys like me were "in what we call the reality-based community," which he defined as people who "believe that solutions emerge from your judicious study of discernible reality." I nodded and murmured something about enlightenment principles and empiricism. He cut me off. "That's not the way the world really works anymore," he continued. "We're an empire now, and when we act, we create our own reality. And while you're studying that reality – judiciously, as you will – we'll act again, creating other new realities, which you can study too, and that's how things will sort out. We're history's actors . . . and you, all of you, will be left to just study what we do."

You see here the basic aggressive tenets of psyops: the idea is not just to distort reality, as propaganda does. It is to transform reality into something that is one's own creation. The masterpiece of psyops in recent times has been the creation of the alleged "Weapons of Mass Destruction" that the government of Iraq was said to stockpile somewhere within the country. It was to those non-existing weapons that Karl Rove was referring when he spoke about "creating reality."

Psyops may also go trans-national and directly target the social and political system of a foreign country. This is a very innovative concept: so far, propaganda had been linked to shared cultural memes in the country where it originated. For instance, during WW2, it was not difficult to convince Americans to hate the "Japs", variously described as evil and monkey-like, but the same techniques would hardly have worked in Japan. Perhaps the first example of a successful transnational psyop may have been with Mata Hari, the Dutch dancer who was accused of espionage and shot by the French in 1917. Not all the details of this story are known, but it seems clear that Mata Hari was not a spy: the case may have been created by the German secret service to balance for the blunder they had made in 1916, when they had shot a British nurse, Edith Cavell, under the same accusation. The allies had amply exploited the Cavell case to paint the Germans as evil Barbarians and the Germans may have just tried to reciprocate [36]. It did not work very well: Mata Hari was amply vilified as an evil *femme fatale* by the French press and her execution did not generate the international indignation that of Edith Cavell had. At that time, psyops were not yet so sophisticated as they are today.

In more recent times, it has been said that the fall of the pro-Russian Ukrainian government in 2014 was the result of a psyop created by the Western Powers in order to bring Ukraine within the Western sphere of influence. The operation went under the name of the "Orange Revolution" and it was just one of the several "color revolutions" taking place in various locations in the world during the past two decades or so, in particular in former Soviet countries, Wikipedia has a list of 23 of them. Some were successful, such as in Ukraine, others have been complete failures, such as the "Violet Revolution" of 2009 aimed at bringing down the prime minister of Italy, Silvio Berlusconi. There is no proof that they were all psyops controlled by foreign powers, but it is possible that at least some were.

Overall, colored revolutions seem to be out of fashion, today, replaced by more sophisticated Web-based operations. The alleged collusion of Donald Trump and the Russian secret services in influencing in the US presidential campaign elections of 2016 is an example of a possible Web-based psyop operation. In 2019, the Special Counsel investigation (also referred to as the Mueller probe or the Mueller investigation) found no evidence of collusion, but it is a safe inference that governments all over the world are involved in trying to affect the policies of other countries. Those who control the Web control the whole world and, at present, the Web seems to be a battlefield where all players in the international arena are engaged in a gigantic struggle.

Psyops do not involve just people wearing colored T-shirts or trolling the internet under false identities. They include targeted assassinations of enemy leaders, false flag operations, terrorism, and more dark and dire things directed against the enemy's government. There is little doubt that psyops have a bright future and the results of the struggle are uncertain but, at least so far, they do not involve human casualties. It is a true "battle of memes" which appear, grow, and then collapse in cyberspace. Where this line of conflict will take us is impossible to say: maybe virtual battles will reduce real violence, or maybe the havoc they wreak will make it worse. As usual, the future cannot be predicted: we need to wait until it becomes the present.

In military matters, there may also exist an "anti-Seneca" strategy. It consists in disregarding Sun Tzu's principle of minimum effort in warfare and aiming instead at continuing the war all the way to the complete military defeat, or even the annihilation, of the enemy. Such a plan could be based on ideological, political, or religious considerations that lead one or both sides to believe that the very existence of the other is a deadly threat that must be removed using force. In ancient times, religious hatred led to the extermination of entire populations and there is a rather well-known statement that may have been pronounced after the fall of the city of Béziers, in Southern France, in 1209. It is said that the Papal legate who was with the attacking Catholic troops was asked what to do with the citizens of Béziers, among whom there surely were Catholics and Albigensian heretics. The answer was "Kill them all, God will know His own." That war, just as most modern wars, was an "identity war" where the enemy is seen as not just an adversary, but an evil entity to be destroyed. These wars tend to be brutal and carried on all the way to the total extermination of the losing side. In some cases, wars may be prolonged because they are good business for some people and companies on both sides.

A possible recent case of this kind of "anti-Seneca" strategy may be found in the campaign that was started in the US in 1914 to provide food for Belgium during the First World War. The campaign is normally described as a great humanitarian success but in the recent book *Prolonging the Agony* (2018) [37], the authors, Docherty and Macgregor, suggest that the relief effort was just the facade for the real task of the operation: supplying food to Germany so that the German army could continue fighting until it was completely destroyed. This seems to be mainly speculation, nevertheless Belgium was occupied by the German army at that time, and so it could be expected that at least part of the food sent there would end up in German hands. But it is also true that, at the time of the campaign, the US was not at war with Germany so the operation can be described simply as a lucrative business for American farmers who found a way to sell food to Germany in this rather indirect way.

Something more ominous took place during the Second World War. By September 1943, after the surrender of Italy, it must have been clear to everybody on both sides that the Allies had won the war, it was only a question of time for them to finish the job. So, what could have prevented the German government from following the example of Italy and surrender, maybe ousting Hitler as the Italian government had done with Mussolini? We do not know whether some members of the German leadership considered this strategy but it seems clear that the Allies did not encourage them. One month after Italy surrendered, in October 1943, Roosevelt, Churchill, and Stalin, signed a document known as the "Moscow Declaration" [38]. Among other things, it stated that:

> At the time of granting of any armistice to any government which may be set up in Germany, those German officers and men and members of the Nazi party who have been responsible for or have taken a consenting part in the above atrocities, massacres and executions will be sent back to the countries in which their abominable deeds were done … and judged on the spot by the peoples whom they have outraged.

> … most assuredly the three Allied powers will pursue them to the uttermost ends of the earth and will deliver them to their accusors in order that justice may be done. … < else > they will be punished by joint decision of the government of the Allies.

What was the purpose of broadcasting this document that threatened the extermination of the German leadership, knowing that it would have been read by the Germans, too? The Allies seemed to want to make sure that the German leaders understood that there was no space for them to negotiate an armistice. The only way out left to the German military was to take the situation in their own hands to get rid of the leaders that the Allied had vowed to punish. That was probably the reason for the assassination attempt carried out against Adolf Hitler on June 20th, 1944. It failed, and we will never know if it would have shortened the war.

Perhaps as a reaction to the attempted assassination of Hitler, on September 21, 1944 the Allies publicly diffused a plan for post-war Germany that had been approved by the British and American governments [39]. The plan, known as the "*Morgenthau Plan*," was proposed by Henry Morgenthau Jr. secretary of the Treasury of the United States. Among other things, it called for the complete destruction of Germany's industrial infrastructure and the transformation of Germany into a purely agricultural society at a nearly Medieval technology level. If carried out as stated, the plan would have killed millions of Germans, since German agriculture, alone, would have been unable to sustain the German population.

Unlike the Moscow declaration that aimed at punishing German leaders, the Morgenthau plan called for the punishment of the whole German population. Again, the proponents must have been aware that their plan was visible to the Germans and that the German government would use it as a propaganda tool. President Roosevelt's son-in-law Lt. Colonel John Boettiger stated that the Morgenthau Plan was "worth thirty divisions to the Germans." [39]. The general upheaval against the plan among the US leadership led President Roosevelt to disavow it. But it may have been one of the reasons that led the Germans to fight to the bitter end.

So, what was the idea behind the Morgenthau plan? As you may imagine, the story generated a number of conspiracy theories. One of these theories proposes that the plan was not conceived by Morgenthau himself, but by his assistant secretary, Harry Dexter White [40]. After the war, White was accused of being a Soviet spy by the Venona investigation, a US counterintelligence effort started during WW2 [41] that was the prelude to the well known "Witch Hunts" carried out by Senator Joseph McCarthy in the 1950s. According to a later interpretation [40], White had acted under instructions from Stalin himself who wanted the Germans to suffer under the Allied occupation so much that they would welcome a Soviet intervention. It goes without saying that this is just speculation but, since this chapter deals with the evil side of collapse, this story fits very well in it.

In the end, there is no evidence that the Morgenthau plan was conceived by evil people gathering in secret in a smoke-filled room. Rather, it has certain logic if examined from the point of view of the people engaged in the war effort against Germany in the 1940s. They had seen Germany rebuilding its army and restarting its war effort to conquer Europe just 20 years after it had been defeated in a way that seemed to be final, in 1918. It is not surprising that they wanted to make sure that it could not happen again. But, according to their experience, it was not sufficient to defeat Germany to obtain that result: no peace treaty, no matter how harsh on the losers, could obtain that. The only way to put to rest forever the German ambitions of conquest was by means of the complete destruction of the German armed forces and the occupation of all Germany. For this, the German forces had to fight like cornered rats and be exterminated. And it seems reasonable that if you want a rat to fight in that way, you have to corner it first. The Morgenthau plan left no hope to the Germans except in terms of a desperate fight to the last man.

We do not know whether the people who conceived the plan saw it in these terms. The documents we have seem to indicate that there was a strong feeling among the people of the American government during the war about the need to punish Germany and the Germans, as described, for instance, in Beschloss's book *The Conquerors* [39]. Whatever the case, fortunately, the Morgenthau plan was never officially adopted and, in 1947, the US changed its focus from destroying Germany to rebuilding it by means of the Marshall plan.

There have been other cases of wars where there was no attempt to apply the wise strategy proposed by Sun-Tzu who suggests to always leave to the enemy a way to escape. Nowadays, wars seem to be becoming more and more polarized and destructive, just as the political debate. And that makes them more destructive: once a war has started, nowadays, the only way to conclude seems to be the complete collapse of the enemy and the extermination of its leaders. The laughter of Hillary Clinton, then US secretary of state, at the news of the death of the leader of Libya, Muammar Gaddafi, in 2011 is a case in point of how brutal and cruel these confrontations have become. It is hard to see how the trend in this direction could be reversed until the current international system of interaction among states that created it collapses. At least, it should be clear that the anti-Seneca strategy is an especially inefficient way to win wars.

To conclude this section on the evil aspects of the Seneca Cliff, we may examine the subject of deception and betrayal as tools to avoid ruin. Lying is surely a very ancient art, can it be used to trigger the collapse of an enemy or of a competitor? On this point, there exists a paradigmatic story: that of the two unarmed men who found themselves facing a hungry lion, somewhere in Africa. While one of the two calmly starts putting on his running shoes, the other asks him, "why are you doing that? Don't you know that the lion can outrun you even if you wear those shoes?" The first man answers, "I don't need to run faster than the lion, I just need to run faster than you."

This story is one of the many narrative versions of the concept that in some conditions one person's gain may be optimized by ensuring another person's loss and that may involve deception and betrayal. In studies on human behavior, collaboration is often the focus [42], but there also exists a scientific literature about betrayal. Much of this work has been done done on the basis of case studies, see for instance the book *Betrayal and Betrayers* by Malin Akerstrom [43]. Another well known method is that of operational games where betrayal is studied in the framework of optimizing the payoff for players in different situations. In this field, you find the "Dictator's Game," the "Ultimatum Game," the "Trust Game," all part of the field known as "Game theory," originally developed by such figures as John Nash and John Von Neumann (see, for instance, the book by Myerson, *Game Theory* [44]). Then, of course, betrayal plays a fundamental role in many competitive boardgames with perhaps the oldest example being *Diplomacy,* a strategic game created by Allan. B. Calhamer in the 1950s. In *Diplomacy*, just as in many strategic boardgames, players take the role of leaders engaged in local or world dominance.

The field of game theory, and of boardgames as well, is vast but we can limit it to those decisions that affect the possibility of a collapse. In other words, when is it convenient to betray someone in order to minimize or avoid one's own collapse? A good example is the well-known "prisoner's game." [45]. This is the way it was described by Poundstone in 1992 [46]

> Two members of a criminal gang are arrested and imprisoned. Each prisoner is in solitary confinement with no means of communicating with the other. The prosecutors lack sufficient evidence to convict the pair on the principal charge, but they have enough to convict both on a lesser charge. Simultaneously, the prosecutors offer each prisoner a bargain. Each prisoner is given the opportunity either to betray the other by testifying that the other committed the crime, or to cooperate with the other by remaining silent.

In the game, betrayal brings a benefit to one of the players only if the other player decides to cooperate. If both defect, they both suffer heavy penalties. And if both cooperate by not betraying the other, they suffer only minor penalties. In principle, the best strategy overall is when players collaborate with each other, but they cannot know what the other will be doing and they may be tempted to defect, hoping that the other will be naive enough to collaborate.

The prisoner's dilemma game has no optimal strategy. Empirical studies show that the simple strategy called "tit for tat" is the one that performs best if the game is played several times with the same players. That is, each player cooperates or defects according to what the other player did in the previous round of the game. In this version, the behavior of a player is based on what he perceives to be the reputation of the other. But there is no guarantee that this strategy will always bring a benefit to those who adopt it. Besides, what to do when playing against someone whose reputation is not known? So, the game reflects the complexity and unpredictability of the real world.

The prisoner's game involves betrayal, but no deception: there is no lying to each other involved. Something similar takes place in the story of the lion and the two men: it involves no deception, either. On the basis of the known data, each player makes a calculation of the odds of two possible strategies: fighting the lion together or running away. There is no real "game" here since there exists an obvious optimal strategy: the man who believes he is faster runs away alone, leaving the slower man to face his personal Seneca cliff in the form of a hungry lion. But, in real life, deception is often a fundamental element of the interaction among human beings.

We may inject deception into the rules of these games. In the story of the lion and the two men, what if only one of the two knows that the lion is coming? This is a version of the game that I called "the camper's dilemma" in 2017 [47]. I described it in terms of a bear threatening two unarmed campers, but the story is the same when it involves a lion or any other dangerous creature. The gist of the game is to decide what is the best strategy to survive when one of the players discovers that a hungry lion, or bear, is near. Is it better to try to survive alone or to cooperate with the other camper? It depends on the situation. Let us imagine that you saw the bear when you were searching for berries while the other camper was near the tent. What you do depends on how serious the threat is (or it is perceived to be). Maybe the bear you saw was far away or maybe it was a small bear, not likely to attack two human beings who fight together. Then, the best strategy is collaboration.

But what if the bear is near and it is a grizzly, so big that you have no hope of surviving a fight, not even if you join forces with your fellow camper? In this case, your best chance of survival is deception. You tell your friend that you will take a walk to collect strawberries and, as soon as you are out of sight, you start running. Your friend will do the same when the grizzly appears, but you have a good advantage and you may be able to survive this mini-stampede.

The "camper's dilemma" game shows that there are situations in which asymmetric knowledge makes betrayal convenient when facing a potential catastrophe. It is a condition that may well apply to real-world situations. Let me give you an example: In 2017, there appeared a piece on "The Guardian" [48] titled, "'We need development': Maldives switches focus from climate threat to mass tourism."

> This week the Maldives, under new president Abdulla Yameen, apparently changed environmental tack, saying that mass tourism and mega-developments rather than solar power and carbon neutrality would enable it to adapt itself to climate change and give its young population hope for the future.

> Fears of immediate sea level rise, which scientists said in the latest IPCC report was accelerating and could mean 75% of the Maldives being under water by 2100, were unfounded, Adam said. "It is not going to happen next year. We have immediate needs. Development must go on, jobs are needed, we have the same aspirations as people in the US or Europe."

As a first impression, these declarations sound like pure madness. The Maldives are islands rising out of the sea on top of coral reefs of no more than a couple of meters on the average. So far, they have been able to survive a sea level rise of the order of centimeters and there is no evidence that they are at immediate risk of sinking [49]. But sea level rising rate is accelerating [50] and for how long will the coral islands be able to cope? Nobody can say for sure, but they may well succumb in a non-remote future since, as far as we know, the islands never experienced the kind of rapid sea level change that global warming is going to generate in the near future [51]. And there is no need for the islands to be completely submerged for their inhabitants to suffer great damage. Coral islands are a very bad place to experience floods: there is no high ground to take refuge on.

So, there are good reasons for the people living on these islands to be worried, but the Maldivian government does not seem to care because it plans to build a *"Riviera-style super-resort with sea sports, six star hotels, high-end housing and several new airports,"* and *"Plans to increase tourism from 1.3 million people a year to more than seven million within 10 years."* Is this a case described by the proverb *"Whom the Gods would destroy they first make mad"*?

The Maldives are not the only archipelago where the local leaders have decided that the threat of global warming should be ignored. Something similar is going on in the Kiribati islands, another archipelago of coral islands in the Pacific Ocean. According to an article which appeared on CBS news [52] in November 2017, the Kiribati government,

> … proclaims the goal of promoting tourism by attracting foreign investors to develop "5-star eco-friendly resorts that would promote world-class diving, fishing and surfing experiences" on currently uninhabited islands. It says the nation's 20-year plan "has an ambitious aim to transform Kiribati into the Dubai or Singapore of the Pacific."

I am sure that the events taking place in the Maldives and in the Kiribati islands remind you of the similar political reversal regarding the policy to face climate change that occurred in the United States in 2016, even though the US is under no threat of being swamped by the waves. More recently, a similar evolution took place in Brazil with the election of Jair Bolsonaro as president in 2019. Among other things, the new president threatened to have Brazil quit the Paris agreement, just like the US did with President Trump.

Why do people start denying the threat as it becomes closer? There may be deep psychological reasons for that, but I might propose a different interpretation. It has to do with the fact that, while at the individual level you can only deceive yourself when facing the Seneca Cliff, at the collective/political level you have the possibility to deceive someone else and, if you are a member of the elite, you may decide to deceive the commoners in order to save yourself.

Here is a recent historical example of the elites deceiving the commoners. In 1943, during the second world war, the Italian high command had been negotiating the surrender of Italy to the Allies for months in complete secrecy. Up to the last moment, the official truth was that there would be no surrender and that the superior fighting spirit of the Italian people would triumph, no matter what the superiority of the Allies was in terms of materials and manpower. Then, when the surrender was made public, on Sep 8, 1943, the King of Italy and the top generals saved themselves by taking refuge with the Allies while the army was left to be "eaten by the lion," in this case the German army.

Now, let us go back to the cases of the Maldivian and the Kiribati archipelagos. Imagine that you are part of the elite of the islands and that you are smart enough to understand what is going on with the Earth's climate. You know that it is unlikely, to say the least, that the people of the rich world would give up their shiny SUVs for the sake of a bunch of wretches living on some remote islands. So, what is the rational thing for you to do? Of course it is to sell what you have and then say good riddance to those who remain. That implies, of course, that you should not tell anyone that you fear that the islands will sink. On the contrary, you must prepare grand plans of development as if you were sure that the islands will stay afloat forever. Then, when things start going bad, you have a chance to leave and join your bank accounts on the mainland. The poor will be stuck where they are, for them, the Seneca Cliff ends underwater.

The cases of small islands are not isolated, only more evident than others. Look at what Donald Trump is doing: he downplays climate change in favor of economic development, just what the Kiribati's and Maldives' governments are doing. If the US elites have decided that there is no hope to save everyone, the logical thing for them is to move into "cheating mode" and let most people die not just by sea level rise, but by starvation, sickness and other consequences of climate change. That gives them the time to prepare, accumulating resources for the coming emergency. Unfortunately, this particular strategy to deal with complex systems under stress has a perverse logic and, if this interpretation is correct, the elites of most of the developed world will soon follow suit in the denial of climate change. We just have to wait and see.

Avoiding Overexploitation. Drill, Baby, Drill!

Fig. 4.4 A pumping jack in an oil field (https://www.publicdomainpictures.net/en/viewimage.php?image=177469&picture=oil-pump-jack)

In 2008, Sarah Palin, then the Republican candidate for the vice-presidency, engaged in a TV debate with her Democratic opponent, Joe Biden. The debate touched on the question of climate change and energy resources with Biden stating that, [53]

> Now, let's look at the facts. We have 3 percent of the world's oil reserves. We consume 25 percent of the oil in the world. John McCain has voted 20 times in the last decade-and-a-half against funding alternative energy sources, clean energy sources, wind, solar, biofuels.

Politicians like to state that they care about facts, except that what they call facts are often more their interpretation of reality than actual reality. But, in this case, Biden was reporting reasonably correct data for 2008 when the "shale boom" of oil production in the US had barely started.

And here is how Sarah Palin answered:

> The chant is 'drill, baby, drill.' And that's what we hear all across this country in our rallies because people are so hungry for those domestic sources of energy to be tapped into. They know that even in my own energy-producing state we

have billions of barrels of oil and hundreds of trillions of cubic feet of clean, green natural gas.

Sarah Palin provided no facts, rather she spoke about a "chant," drill, baby, drill," a magic spell, an enchantment, an exorcism. In terms of facts, she provided only vague estimates using resounding words in terms of "billions of barrels" and "trillions of cubic feet."

This is the way politics works: using magic rather than facts to convince people. It is all part of an ongoing trend in politics: over the past decades the political discourse has become more emotional and less fact-based, pivoting around the capability of the big man at the top (rarely the big woman) to sound convinced and reassuring. It is a trend that's described in a recent paper by Jordan et al. [54] as

> Across multiple corpora from the American presidents, non-US leaders, and legislative bodies spanning decades, there has been a general decline in analytic thinking and a rise in confidence in most political contexts, with the largest and most consistent changes found in the American presidency.

The Palin/McCain team was defeated by the Biden/Obama team in 2008 but that changed little to the fact that Palin's proposal to drill more overcame Biden's idea of moving to renewables. In politics, one of the main rules for success is "*all changes you propose must have the purpose of avoiding change.*" Biden was proposing to move to clean energy: that meant real change and that is a no-no in politics. Palin was proposing no change at all, except maybe chanting some mantra all together. That was a winning strategy in political terms. Fortunately for the Obama/Biden team, climate change and energy remained marginal themes in the debate.

The idea of drilling more was already in motion before the 2008 election and it progressively gained ground. The financial world provided resources for the industry to engage in a major effort to extract more oil and that could be done by exploiting from shale deposits. It is a kind of oil contained in the rock matrix in bubbles not interconnected with each other, so that the gas or the liquid cannot spontaneously flow to the surface once the rock is drilled. To get the oil, it is necessary to create a path for the oil to flow by fracturing the rock (or "fracking," as it became fashionable to say in recent times). In the old times of the petroleum industry, it is said that it could be done by throwing a lighted dynamite stick into the borehole, nowadays it is done by injecting high-pressure fluids inside the rock. It does not change the basic idea so much, although the dynamite stick was probably more spectacular.

Despite the complexity and the high cost of fracking, in a few years the US oil industry managed to invert the declining trend that had been ongoing from the 1970s. With the 2010s, drilling increasingly became the accepted wisdom while renewable energy was gradually going out of fashion or relegated to some marginal regions of the debate while most politicians engaged in new magic slogans such as "clean coal" and "green growth." The "drill, baby, drill" chant triumphed and the oil depletion problem seemed to have been pushed to a future so remote that nobody would have to worry about that anymore. In time, Palin's 2008 chant of "drill, baby, drill" was transmogrified into the one called today "energy dominance," another magic slogan used for the first time by Donald Trump in 2017. An interesting concept: it is as if you could dominate your neighbors by burning your house faster than they are doing. But never mind the logic of that: aren't we dealing with magic?

Extracting shale oil may be described as "magic" by politicians, but surely it is a complex and expensive technology. To give you some idea of the difficulties involved, note how a recent article from China by Stephen Chen [55] reports how nuclear weapon technologies could be used to mobilize hydrocarbons trapped in shale deposits. Not that the plan is to detonate nuclear warheads for that purpose, but the device described in the article is called an "energy rod" able to create shock waves that will fracture the underground rock. Apart from sounding a little like the staff of Gandalf the White in Tolkien's trilogy, it seems to be an especially expensive and complicated variant of the old idea of dropping dynamite sticks into the borehole. Given the costs and the difficulties involved, we cannot say for how long the shale boom will last. What we can say is that, so far, the shale industry has not provided much of a profit for investors [56]. So, for how long can the industry keep going like that? The Seneca Cliff for the shale oil industry may not be far away in the future. In politics, magic always wins against reality— but only for a while.

The Palin versus Biden debate is a good starting point to discuss a very general question: how should we manage the Earth's natural resources? Can we really keep growing forever, as most politicians seem to imply? Or do we face the Seneca Cliff for our whole civilization when we start truly running out of the resources that created it?

All natural resources are scarce by definition: if they were not, they would come for free. This is why you do not pay for the oxygen you breathe nor for the sunlight coming through the window (so far, at least). But oil, gas, gold, whales, grain, and caribou are all examples of limited resources, a well-known concept in economics. Economists normally agree on a concept called "general equilibrium theory" which implies that if demand exceeds production, prices will rise, reducing the demand and/or generating new investments that will increase production. In both cases, equilibrium will be restored. The opposite will take place if production exceeds the demand.

These concepts are considered proven within the assumptions at the basis of modern economics, but are they true in the real world? Kate Raworth notes in her book *Doughnut Economics* (2017) how the early economists banked on Newton's prestige to make economics "laws" look like physical laws, similar to the laws governing the motion of planets. Raworth remarks (p. 135)

> One thing that's clearly coming to an end is the credibility of general equilibrium economics. Its metaphors and models were devised to mimic Newtonian mechanics, but the pendulum of prices, the market mechanisms, and the reliable return to rest are simply not suited to understanding the economy's behavior. Why not? It is just the wrong kind of science.

Raworth means that Newtonian mechanics is perfectly suitable to describe the motions of bodies in a gravitational field as an approach that naturally leads to a condition of equilibrium. But the economic system is not in equilibrium. It may be in homeostasis—a condition that may look like equilibrium, but that is a completely different concept. The market is well known to go through cycles of growth and decline and prices normally oscillate, sometimes wildly, something equilibrium physics cannot describe. Physics and economics stand to each other a little like chess and paintball, they are both games simulating real battles, but with very different rules.

The problem is most evident when we discuss non-renewable resources. When Sarah Palin was promoting her "drill, baby, drill" chant, she meant that every oil company should strive to maximize both production and profits. But if oil is a non-renewable resource, then drilling more will only lead to run out of it faster—although operators may be able to enjoy the short-lived abundance. The reason why depletion was neglected in the debate is due in large part to the human tendency to discount the future, in other words to think that an egg today is better than a chicken in the future. This is a big problem and it seems that, for most people, events that are expected to occur more than about five years in the future are just not considered important.

Nevertheless, economists do not just tell people, "eat your egg as long as you have it." On the contrary, already about one century ago, economists started thinking about the problem of depletion. The basic idea that seems to be still current in this field is that the efficiency of the market in allocating scarce resources should be able to take care of optimizing the exploitation of non-renewable ones as well. So, as producers deplete a stock, the all-knowing market will perceive the increasing scarcity and react by increasing the price of the product. That allows producers to maintain their production despite the higher costs while seeking for new resources which could be of the same kind, but more expensive to produce, or completely different ones, possibly renewable ones. According to a model developed for the first time by Harold Hotelling in 1931 [57], the result will be a smooth substitution of the depleted resource with a new one, called the "backstop resource."

You may object that it is an act of faith that there will be always something available to replace a resource that has become too expensive to be used. Indeed, in many cases the belief of the availability of replacements is built on a rather naive faith in technological progress. But it is also true that many scarce resources can be replaced with less scarce ones. Over the past few centuries, coal replaced wood, oil replaced coal, natural gas may be replacing oil. And we can replace copper with aluminum, zinc with titanium, plastic with bio-plastic, and so on.

This line of reasoning has led to some overoptimistic assessments in the past, such as the "principle of infinite substitutability" proposed in 1978 by Goeller and Weinberg [58], mainly based on what appeared to be a promise of cheap and abundant energy obtainable from nuclear power. We tend to be less optimistic, nowadays, but it is also true that physical scarcity, in itself, is not an unsolvable problem: replacement, recycling, efficiency, restructuring, are all strategies that can be used to fight the depletion of mineral resources. After all, humans can hardly mine themselves out: everything we extracted in the past has not disappeared, it is somewhere and will remain forever with us. So, nothing prevents us from using the same strategy that has been used by plants to "mine" the crust for hundreds of millions of years without ever running out of anything. How did they manage that? On the basis of three fundamental principles: (1) use only what is abundant, (2) use as little as possible, (3) recycle ferociously.

It worked for plants and it is still working for the whole biosphere, but could we do the same with our industrial system? Not easy, of course, but there are no physical reasons why it could not be done. Some people have a wrong understanding of the second principle of thermodynamics and assume that because entropy is supposed to increase always, then it will never be possible to completely recycle minerals. But the second principle works only for isolated systems and our planet is not one—that is why plants could manage to recycle everything for so long. The problem with recycling is not thermodynamics, but the cost, and it is hard to think that the deity called "free market" will do the miracle for us with no pain involved. Moving to 100% recycling involves forsaking the current "energy subsidy" that millions of years of sunlight and other forces have accumulated in mineral ores—we'll have to pay the price for this energy ourselves and that implies a complete rethinking of the way we extract, use, and recycle mineral. A change of attitude that looks very unlikely considering that the government of the US seems to have fully embraced the idea that the way to deal with oil depletion is to extract what is left at the fastest possible speed in the largest amount possible, without thinking—even vaguely—of the necessity of investing in a replacement for the future. We have a lot to learn in this field.

But what about renewable resources? In principle, we can keep producing biological resources—wood, grain, food, fiber, and more—as long as there is sunlight to power the photosynthesis process, can't we? Unfortunately, we do have a problem of depletion also with renewable resources, a problem that can be even worse than that with the non-renewable ones. Human beings are so good at exploiting resources that they tend to destroy them, creating a scarcity that, in itself, would not need to exist.

It is a story that goes back to very ancient times. Think of how American Indians used to kill bison by pushing them down a cliff and making sure that not a single one survived, as told by Lewis and Clark in the report of their 1804–186 expedition [59]. The idea that the best way to get a bison steak for dinner is to exterminate a whole herd does not seem to be the most efficient one, but this attitude may have been typical of our remote ancestors. Indeed humans are often accused of having been the cause of the pulse of extinctions of "megafauna" (creatures weighing more than 100 lbs or 44 kg) observed around 10,000 years ago [60]. This is a controversial point and there are other possible causes for ancient extinctions, but it is also true that we have direct historical evidence of how modern wasteful hunting practices led to the near—or total—extinction of large animals. If you read Melville's *Moby Dick*, you surely noticed how 19th century whalers would kill whales to get just a few liters of the *spermaceti* oil contained in their large heads, the rest they would throw away except for a few chunks, such as when we read of first mate Starbuck eating a whale steak on the deck of the *Pequod*. From the age of whaling, things have not changed so much and we have not really learned how to manage the exploitation of marine creatures. Having nearly run out of several species of whales [61], we now risk running out of much smaller creatures, such as squid [62].

Why do people keep destroying the resources that make them live? Gandhi is reported to have said that "the Earth provides enough to satisfy every man's need, but not every man's greed." This statement can be understood not as meaning that humans can expand their numbers forever but that an economic system based on greed will always create needs that the Earth will not be able to satisfy. Unfortunately, the idea that greed is good is enshrined in current economic thought and economists seem to have been slow in detecting the gaping hole at the basis of their views.

That's exactly where the problem lies: it is called "overshoot" and we saw its description in an earlier chapter of this book. The more you go in overshoot, the harder you have to "return" to a flow rate well below the carrying capacity of the system. Unfortunately, the tendency of a system that works simply according to maximizing dissipation of the resources it uses is equivalent to maximizing the utility function of the operators: nobody is in control except for the abstract entity we may call "Greed." It is like following Sarah Palin's suggestion in the form, "exploit, baby, exploit." In all fields, everyone tries to maximize production and the result is a rollercoaster economy. And, at times, the rollercoaster may well crash into the ground when a resource is exploited to a level below its capability to rebuild itself. In biological systems, extinction is forever.

These problems are generally recognized nowadays, even though not always expressed in a form that takes into account the dynamic factors of overshoot and collapse. The way to solve them has normally been to emphasize individual commitment and goodwill. A good citizen, it is said, participates in the fight against climate change by consuming less and polluting less than what is imposed on him or her by law. It is a very common idea: there are few discussions on climate change and pollution that do not end with a brief list of recommendations, such as to use bikes, turn off the lights when one is not at home, buy groceries from local producers, use natural fibers, and the like. It is not even a new idea, the Stoics at the time of

Seneca were doing the same: faced with a terrible dictatorial government they had no power to control, they emphasized personal virtue and, yes, "stoicism" against the unavoidable adversities.

But can individual goodwill avoid the overexploitation of natural resources? Despite all efforts done up to now, it is hard to think that drinking your coke without using a plastic straw will do anything significant to solve our environmental problems. The problem is simple: a person's restraint is another person's opportunity. In other words, a person who is a good ecologist and decides to go to work by bike may simply free fuel resources that a less conscientious person may use to go to work on an SUV. It is something similar, but slightly different from Jevons' paradox. It is what I called the "hummingbird effect" [63]. The idea comes from the old story of a hummingbird trying to extinguish a giant forest fire while carrying just a drop of water in its beak. It is, of course, useless against the fire, but the hummingbird is very proud of what he is doing and, in the story, the little bird is praised for his willingness to do its duty against all odds. Humans, it seems, have a similar attitude: they tend to be very proud of some minor contributions against global warming they engage in, say not using plastic straws for their drink, but using several tons of fossil fuels for their summer vacations. Jean Baptiste Comby described the problem in his 2015 book *La question climatique* ("The Climate Question") [64]. He didn't use the hummingbird analogy but he argued that the climate question has been thoroughly depoliticized and consigned wholly to the realm of individual decisions. A way to make people feel good, but with little or no impact on the system.

It seems that it starts being recognized, today, that individual actions are insufficient to solve the problems we are facing and avoid the impending climate and depletion cliff. That is the reason for the appearance of such political movements as the "Extinction Rebellion," emphasizing collective action. A popular leader in this field has been the young Swedish activist Greta Thunberg. Her action is clearly framed in collective terms: her message rarely includes recommendations on individual actions such as "don't take a plane if you can get there by train" (although she does that, too). She speaks to leaders asking them to do something to ensure that the people of her generation will have a future. It is clear in her message that this action will carry a cost that most of us will have to pay. Will this message be heard, or will the environmental movement continue to toy with double pane windows?

Controlling of Complex Systems: The Story of the Last Roman Empress

Fig. 4.5 This is perhaps the only realistic portrait we have of Galla Placidia (388–450 c. e.), the last (and the only) Western Roman Empress. The inscription says "Domina Nostra, Galla Placidia, Pia, Felix, Augusta," that is "Our Lady, Galla Placidia, Pious, Blessed and Venerable." A contemporary of such figures as Saint Augustine, Saint Patrick, Attila the Hun, and—perhaps—King Arthur, Placidia had the rare chance of being able to do something that past Roman Emperors never could do; take the Empire to its next stage which was to be, unavoidably, its demise (Image by Clio20—https://en. wikipedia.org/wiki/Galla_Placidia#/media/File:Honorius_et_Galla_Placidia.JPG)

The story of Galla Placidia reads like an adventure novel [65]. Born in late 4th century CE, she lived most of her life during the last century of the Western Roman Empire. In 410 CE, she was a young Roman princess when she was kidnapped by the Goths during the sack of Rome. Undeterred, she married their king and became their queen. There followed more dramatic events: her husband, the king of the Goths, was killed in a conspiracy and Placidia went back to Roman lands, battling against her half-brother, Honorius, for the Imperial throne in the city of Ravenna, at that time the capital of the Western Empire. Defeated, Placidia had to flee, but Honorius died and she came back at the head of an army to retake Ravenna, in the meantime occupied by a usurper. Placidia defeated the usurper, captured him, had his hand cut off, paraded him in town riding a donkey, and finally had him beheaded. In 425 CE, the victorious Placidia took for herself, alone, the title of *Augusta* (venerable) that had belonged to the first Roman Emperor, Julius Caesar, some 500 years before her (Fig. 4.5).

As I said, Placidia's story is truly an adventure novel and it is strange that nobody ever thought of turning it into a movie. After all, Placidia was a contemporary of such well-known figures as Attila the Hun and, (perhaps) King Arthur of Britain, both much more popular than her in fiction. But the interest in Placidia's life and deeds is not limited to her juvenile adventures. As Empress, she never was just a doll in expensive clothes. Rather, she was possibly the last person who actually ruled the Empire: she faced enormous problems but managed to keep the Empire together. After her death in 450 CE, no one was left who could do the same and the Empire faded away forever.

I can imagine that, at times, many of us dreamed of being what Placidia had managed to become: the absolute ruler of the world. I am sure we all have in our mind the perfect recipe for solving the world's problems: hunger, wars, pollution, global warming, and more—it would surely work if only we had the power to impose our ideas as benevolent and merciful rulers. That's just a dream, of course, but it is true that the Roman Emperors were powerful, semi-divine rulers. They were said to be people "born in the purple," indicating that from childhood they would wear clothes dyed with purple made in Tyre, so expensive to produce that it was reserved for kings and emperors. But then, suppose you were one of those purple-wearing emperors, what would you do to save a collapsing empire?

In general, the record of the performance of Roman Emperors is terribly poor. We all know of Emperor Nero who was accused of having set Rome on fire to find inspiration for one of his songs, and of Caligula who nominated his horse as a senator and engaged in all sorts of debaucheries. Probably much in these accusations is legend and propaganda, but it is true that absolute rulers are often psychologically unstable individuals: they may be murderers, sexual predators, sadists, and worse than that. Even when they succeed in maintaining a certain level of mental sanity, the task of managing a whole state is beyond the capabilities of a single person. To be effective, rulers need competent staff to inform them and guide their decisions, but they tend to surround themselves with yes-men who amplify their biases and misconceptions. Absolute rulers do not solve problems, they *are* problems.

Curiously, there seems to be an exception to this rule: Galla Placidia. She may have been a rare case of a ruler who understood what was wrong in the system and acted accordingly. At the time of Galla Placidia, the last century of the Western Roman Empire, the problem for the Roman state was mainly financial: with the gold mines of Spain exhausted, the Empire had run out of money. In other words, the Empire was in full financial overshoot: it was spending more than it could earn. The previous Roman Emperors had tried to refill the imperial coffers by increasing taxes—but that meant straining the system, making it more fragile. The more they raised taxes to be spent on more troops, the poorer the Empire became, less and less able to face the Barbarian invasions.

Instead, Placidia did exactly the opposite. For sure, she didn't think that wars were a good way to solve the Empire's problems. Cassiodorus (c. 485—c. 585) described her ruling years as involving "too much peace," even though it was intended as a criticism. Stewart Oost, who wrote Placidia's biography in 1969 [66], reports that she enacted two especially interesting laws. One forbade the *coloni*, the peasants bound to the land, to enlist in the army. That deprived the army of one of its sources of manpower and we may imagine that it greatly weakened it. The other law allowed the great landowners to tax their subjects themselves. This deprived the Imperial Court of its main source of revenues and it surely forced the Court to reduce its expenses. These two laws were the push needed to gently nudge the Empire toward its next stage: its demise.

Did Placidia understand what she was doing? Of course, we have no way to know the inner thoughts of a person who lived a millennium and a half before us and who left us nothing written by herself. But she must have been steeped in the ways of seeing the world that were typical of late antiquity in Europe, including a strong influence from Stoic philosophy. In addition, she had lived with the Goths, she could probably speak their language, and she never reneged the title of queen that she had gained with them. That experience may have opened her mind and made her think in ways that were different from the narrow views that we can imagine are typical of a cloistered emperor or empress. So, she applied a strategy consisting in not opposing the unavoidable. Placidia did not try to push the system in a direction where it could not go and she played a fundamental role in opening the way for the coming of the Middle Ages.

This Excursus in Roman history is an introduction to the concept of the control of complex systems. In general, human societies, living creatures, human-made devices, and other kinds of complex systems tend to reach a specific state—sometimes called "homeostasis"—and to maintain it. In some cases, this is the result of the interaction among the internal feedback mechanisms of the system which tend to balance each other. A good example is a flock of birds. The flock is kept together by feedback-dominated inter-actions among single birds. It has no structure that we could identify as a control system: no "Emperor bird" at the top gives orders to the other birds!

Instead, some complex systems have structures specifically dedicated to control. The nervous system and the brain of vertebrates is an obvious example. Another one is a 19th-century invention that made it possible to run steam engines in a reliable manner: the "steam governor," an automatic valve to regulate the flow of steam into the engine (Fig. 4.6). The steam governor was the precursor of the modern concept of control systems for our machines and devices: many are simply set point systems, just like the thermostat that regulates the temperature of a room. Others can actively chase a moving set point, like an automatic anti-aircraft gun. And some can be very complex and adaptive, you can think of the control mechanism that keeps a flying drone stable despite the various maneuvers it performs. The latest example of how sophisticated these systems can become is the currently very fashionable self-driving car, expected to revolutionize road transportation.

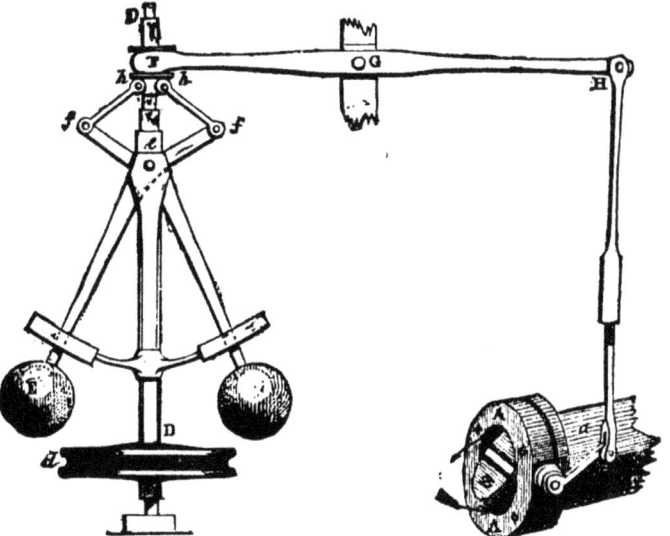

FIG. 4.—*Governor and Throttle-Valve.*

Fig. 4.6   The Centrifugal Steam Governor: an early automatic control device to regulate the flow of steam into the engine. It was the precursor of all modern control devices. Image from *"Discoveries & Inventions of the Nineteenth Century"* by R. Routledge, 13th edition, 1900. https://en.wikipedia.org/wiki/Centrifugal_governor#/media/File:Centrifugal_governor.png

The steam governor greatly impressed the scientists of the 19th century with its capabilities that, up to then, had been thought to be characteristic of living beings only. By means of its internal feedback-based control system, you could see the governor as endowed with a certain degree of "intelligence," reacting to changes in its environment, adapting to new conditions. Similar capabilities exist for living beings: your body, for instance, is a tangle of feedback-based control systems. The level of sugar in the blood is controlled by the synthesis of the insulin hormone. The body temperature is controlled by neural feedback mechanisms operated by the hypothalamus gland, which also contains temperature sensors. And the blood pressure is controlled by a system called the Renin-Angiotensin-Aldosterone System (RAAS). All these systems may malfunction, that is why you may have to take blood pressure control pills. Or, the set point may be varied depending on circumstances, such as when your body temperature increases as a response to infection: it is called "fever." The most basic control system of your body is the one that prevents your cells from growing and reproducing at the fastest possible speed. If that system ceases to work, the result is called cancer.

But not all complex systems have control mechanisms that can keep them in homeostasis. For instance, there is no set point for populations in ecosystems: amoebas in a Petri dish reproduce to increase their numbers as fast as possible and the total is kept in check only by the limited availability of food. It is no different for vertebrate populations: there are no set limits except the one generated by the availability of food. There is a logic in all this: individual creatures have internal set-points and control mechanisms because that makes them better at competing for survival. But there is little or no reason why these mechanisms should have evolved at the group or at the species level, so there is none. Only "eusocial" species, ants, for instance, actively control their population.

For human societies, it does not seem that there exist biological control mechanisms limiting, for instance, population or resource exploitation. But it is also true that we are a partly eusocial species and that we have developed cultural mechanisms supposed to reduce individual independence for the benefit of the community. They take the form of laws, religions, social rules, and more. Many human social structures rely on some kind of "central processing unit" that may go under various names: boss, chief, commander, king, emperor, or—more simply–the "government."

Governments have many purposes, but the overall impression is that they exist mainly to harass their citizens with more and more taxes in order to maintain themselves. Apart from that, all over history governments have tended to justify their existence in terms of defending their citizens from (sometimes real) threats: crime, terrorism, foreign invasions, and the like. Only in relatively recent times, has it become commonplace to believe that the government had to intervene in the economy in ways other than simply issuing currency. An extreme view in this field is that all the means of production should be owned by the state and controlled by the government in order to avoid the waste that is generated by competition among different producers. This view is typical of socialism, but it has been largely abandoned today. Yet, it is still believed that when the economy does not work as it should, the government should do something.

But what should a government exactly do? Financial matters are the most debated area of government action and they can be seen as attempts to control the system by acting, for instance, on the interest rate. The problem is that, here as in other sectors, the government is not normally trying to control the economy in the sense of stabilizing it. Rather, it tries its best to make it grow at the fastest possible speed. For most people, this is supposed to be the obvious thing to do but it may not be such a smart idea. It is as if the governor of a steam engine were to be operated to open the valve as much as possible, all the time. That could lead the machine to rev up over its limits and maybe even explode.

We saw in a previous section how the attempt to keep the flow of natural resources growing, the "drill, baby, drill" approach, has similar consequences. It sends the system in overshoot and then causes it to crash down generating what we call here the "Seneca Cliff." Individual operators or single firms are perfectly capable of generating a collapse by resource overexploitation, but it is an especially destructive effect when several operators or firms compete for the same resource. In that case, the operator who shows restraint and tries to avoid going in overshoot would simply leave more of the resource to another, less scrupulous, operator.

It was a biologist, Garret Hardin (1915–2003), who first noted how the economy was subjected to this problem when he published a famous paper in "Science" in 1968, titled the "*The Tragedy of the Commons* [67]." Hardin's model is the same as the one by Lotka and Volterra that we saw earlier on in this book, except that it was expressed in words rather than using differential equations. Hardin proposed a model based on a hypothetical pasture managed as a "commons," that is, free for everyone, where a number of shepherds could bring their sheep. Shepherd will tend to increase the size of their flocks to increase their profit and that will result in overgrazing. That is, grass will be eaten by the sheep faster than it can grow back. The sheep will starve and the shepherds will see their flocks collapsing. And there comes the Seneca Cliff.

There is little evidence that Hardin's tragedy of the commons actually takes place in pastures [68]. But it was found later on that the Hardin model does describe some economic systems, such as fisheries [69] just as Volterra's studies had demonstrated earlier on [70]. Hardin had identified what we call today the problem of overshoot and collapse, although he did not use these terms in his papers. His ideas were revolutionary in the sense that they showed that in some conditions economic systems do *not* tend to reach the situation of stability that the general equilibrium theory assumes they should when left alone in conditions of "perfect" free markets. Hardin's model was much discussed and often rejected, but it has been lingering in the debate on how to manage the economy.

In parallel with Hardin's considerations, the question of overshoot and collapse was being examined within the new approaches to complex systems. Jay Forrester, the founder of system dynamics was probably the first to use this terminology, noting how economic and biological systems tend to behave like electronic circuits when they "overshoot" the signal and then "return" in a series of damped oscillations [71]. This led Forrester to the first dynamic study of the world's economic system, published in 1971 [72] and his coworkers to the other milestone study *The Limits to Growth* of 1972 [10]. These studies went beyond the hypothetical pastures that Hardin had used as a metaphor and used real-world data to study the world's economy. The result was that the global economy was—or would soon be—in overshoot and that it would have had to return below the carrying capacity of the world's system. This return would be painful, to say the least. Neither Forrester nor the authors of *The Limits to Growth* used the term "Seneca Collapse" but that was what they had identified for the first time in the story of dynamic modeling.

Forrester and the authors of the *Limits to Growth* did not just recognize the problem, they proposed solutions for it. If you want to avoid the overexploitation of a natural resource, then you have to regulate its flow so that the throughput of the exploitation does not exceed the carrying capacity of the system. [73] Both studies showed how the phenomenon of overshoot and collapse could be avoided by putting brakes on some of the main elements of the economic system: the exploitation of natural resources should be slowed down, the human population growth should be stopped, increasing amounts of resources should be dedicated to fighting pollution. The result of implementing these policies was that the world's economy would not go in overshoot and then collapse but would reach a steady state condition that could be maintained throughout the 21st century, at least (Fig. 4.7).

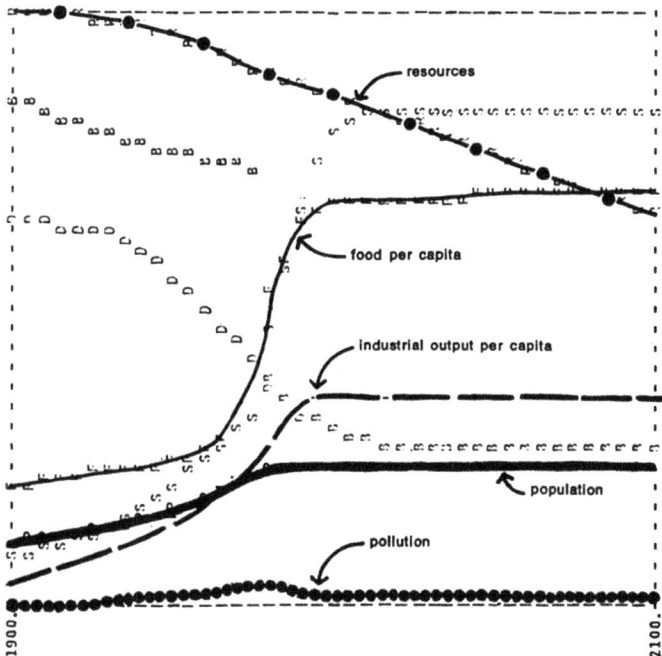

Fig. 4.7 One of the "stabilizing" scenarios proposed in the 1972 The Limits to Growth study. It assumes that the growth of some sectors of the economy is curbed starting in 1975 (Right to reproduce courtesy by copyright owner, Mr. Dennis Meadows)

These results were obtained considering the world's whole economy but they are valid also for smaller economies at the level of single states. The authors also never exactly specified what kind of entity should implement the proposed stabilization policies, but it seems obvious that it could have been only some form of government. Basically, avoiding disastrous phenomena of overshoot and collapse required the government to operate in a way not so different from that of the governor of a steam engine (and, indeed, their name is almost the same!). A governor regulates the speed of rotation of the engine to a predefined set-point, preventing it from running so fast that it could damage itself. A government should do the same, regulating the flow of natural resources into the economy and managing the output in such a way that the "engine"—or the whole society—runs smoothly, avoiding the overexploitation trap.

But we have a problem, here. Whereas centrifugal governors have an excellent record of being able to control steam engines, governments don't enjoy the same good reputation. If you ever tried to push your government to do something sane that would benefit everybody, you understand what seems to be a general rule. A government is nothing like a thermostat or the governor of a steam engine. It is, rather, the embodiment of the concept of the tragedy of the commons described by Hardin with all the actors (lobbies) pushing to grab what they can, when they can, for themselves.

Today, in the West we tend to believe that liberal democracy is the best system of government and, for sure, it has several good points. But it is clearly unable to avoid the overexploitation of the commons. It seems to be a built-in feature: in a democracy, a politician who implements laws that require citizens to make sacrifices to reduce their consumption is not re-elected. The result is that there is no Western leader, at present, who can afford to declare that economic growth may not be the one and the only way to take us toward the nirvana of ever-lasting growth: the best of all possible worlds.

Maybe democracy is not such a great idea, surely not so good to be worth exporting by means of aerial bombing of the unfortunates who do not have it. Among others, the concept that we need different political systems has been expressed by Jorgen Randers [74], one of the authors of the first *The Limits to Growth* report [10]. Randers does not advocate dictatorship, but he thinks we should learn from China how a government should act forcefully when necessary, even against the opposition of the population. The one "one-child" policy enacted by the Chinese government starting in 1979 is a rare example of a successful quota imposed by a government.

The growing opinion that democracy is unable to face the challenges ahead may be a factor in the trend of more authoritarian governments appearing in the West, often with a focus on a single, powerful figure at the head. Yet, it does not seem that the new big men at the top are doing any better than the old parliament-based democracies in terms of protecting the ecosystem. The cases of Jair Bolsonaro, president of Brazil, and of Donald Trump, president of the United States, are clear evidence of this trend: both are heavily focused on promoting economic growth and engaged in dismantling the rules to protect the ecosystem conceived by previous governments. Some leaders, such as Emmanuel Macron in France, claim to be in favor of environmental policies but that seems to to be mainly a veneer of "green" painted over a traditional approach. In practice, the world governments continue to engage in their traditional power games, competing in terms of spheres of influence and occasionally waging wars on each other. Nobody in charge seems to understand that the problem, nowadays, is not that of expanding their country's borders but to ensure the physical survival of their citizens from potentially disastrous events related to climate change and the collapse of the ecosystem.

So bad is the record of many governments nowadays that some people arrived at the conclusion that the only good government is no government at all (just like the only good Indian was no Indian in the views of some 19th century Americans). One result is the extreme Libertarianism of some sectors of the political right in the US, from where there comes the idea that the economic system should be left absolutely and completely free to regulate itself. But if that is the solution, how to avoid the tragedy of the commons? The Libertarian answer to the question is privatization. If every economic actor owns a slice of the resource being exploited, then they won't have any interest in overexploiting it. It has been suggested that the wave of privatizations that swept the world during the past decades was a direct result of Hardin's ideas or, at least, of how they were understood in some political sectors [75] (But note that Hardin himself never advocated privatization.)

At first sight, privatizing the commons seems to be a good idea. Surely, greed is a powerful force in determining people's behavior, so why not exploit it to avoid overshoot? But things are not so simple. One problem is that people may well overexploit resources that they completely control, as appears from a series of studies carried out by Erwin Moxnes [76] that show how people easily misjudge the amount of resources available and the capability of the system to recover after having been perturbed. Jay Forrester also examined this problem with the model he called the "Beer Game" where he showed how managers can completely lose control of a system even when they have the right data and the full capability of acting on it [77]. That may not be a critical problem: people do make mistakes, but they can also learn from them. The real problem with the idea of privatizing the commons is that it does not mean that you do not need a government. For middle-class Westerners, private property may appear an obvious feature of their world: they expect their governments to guarantee their property rights. But this is not true in many areas of the world where ordinary people are subjected to be evicted, dispossessed, or worse. There is a long series of cases in history of entire peoples being chased away from lands they thought they owned; the classic case being that of the American Indians in the 19th century. And, everywhere in old times, property rights were not guaranteed by anyone except by the capability of the owner to defend it using arms. But that is hardly a good way to organize the exploitation of natural resources. If nothing else, it invites the most powerful players in the economic game to behave as pirates, using force to dispossess the weaker ones. Besides, in many cases privatization is simply impossible: for instance, you cannot fence the ocean to prevent fishermen from destroying entire fisheries. Even more difficult would be to use this strategy to manage climate change by privatizing the atmosphere.

So, it seems that we do need some kind of a government but, if the current forms of democracy are unable to carry out the task of stabilizing the economy, could we think of different kinds of political systems? Many ambitious utopias have been proposed in the past, starting from Plato's *Republic*, written around 380 BC. Plato's ideas were never put into practice but during the past few centuries the trend of experimenting with new political theories seems to have become frantic. We had Socialism, Communism, Fascism, Nazism, and more ideologies that were supposed to be at the basis of governments that could take forms such as monarchy, aristocracy, plutocracy, oligarchy, democracy, theocracy, tyranny, and more.

The results have been variable, in most cases very bad. It seems that many revolutionary movements start with noble and lofty ideas on how to reform the government and turn it into something that would work in the name of "we, the people," as in the US constitution. In practice, all political systems tend to degenerate: they may become ineffective kleptocracies, hideous dictatorships, or other forms that just create misery and disasters for everybody. And if you think that Capitalism is the big bad wolf of the story you just have to think of how the government of the Soviet Union destroyed the ecosystem of the Aral Sea to understand that Communism, theoretically the bugaboo of Capitalism, is not a solution for the overexploitation problem (at least in the Soviet version).

Does that mean we are condemned to an eternal series of cycles of growth and collapse as wolves and foxes experience in the simplified Lotka–Volterra model? Or, like in the Buddhist view, can we escape the cycle of death and reincarnation and attain the Nirvana of sustainability? These are difficult questions but, as Thomas Browne said, even the song that sirens sang is not beyond all conjectures.

One thing that is sure is that speculating about political systems may be dangerous. Over history, there have been several cases of people trying to put someone else's political speculations into practice: the result has often been major disasters, as we all know. Instead, we may do better if we look for historical examples of governments that did succeed in managing the commons without having to oppress their citizens (not too much at least). At least one such example exists: Japan during the Edo period, from 1603 to 1868.

The Edo period in Japan is also known as the "Tokugawa period" and it started when the warlord Tokugawa Ieyasu managed to end the age of civil wars (the *Sengoku jidai*) and unify Japan under a military government called the *bafuku,* headed by a commander in chief called the "*Shōgun.*" It is a period that, in the West, we mainly know because of the many Samurai movies that use that period as a setting. But having been a battleground for swordmasters is not the main reason of interest of the Edo period, rather, we can examine it as a relatively recent example of a true "zero-growth" society.

We have no data about Edo Japan that we could directly compare to our modern concept of "Gross Domestic Product," at the basis of our idea of economic growth, but we know that the Japanese economy was lively and growing in terms of wealth per capita [78]. Remarkably, this economic growth did not result in an increasing population. After an initial period of expansion, from ca. 1700 onward, the Japanese population stabilized to a level of around 26–27 million people [79], a number that remained unchanged until 1854, when Commodore Perry used his "Black Ships" as shock and awe tools to force Japan out of its economic isolation to restart a period of expansion. We also know that the extent of cultivated land in Japan did not vary over almost one century and a half, from 1720 to 1874 [80]. We have some records of famines during this period, but they seem to have been rare and related to special climatic events, such as volcanic eruptions. Overall, we can say that for some two centuries Japan was as close to a "zero-growth" society as we can imagine one.

How did Japan manage to attain this condition? Probably, the simplest answer is that the Japanese had no other choice. They had tried military expansion under the leadership of the warlord Toyotomi Hideyoshi who launched two offensives against Korea in 1592 and in 1597, but the effort was not successful and that forced the Japanese to face the necessity of living within the limits of their islands.

But how was zero growth obtained? First of all, it does not seem that the government had a plan to ensure the sustainability of the Japanese economy. Like most governments in history, the *bafuku* was mostly interested in its own survival. For this purpose, it implemented a strict control over all the sectors of the Japanese society by means of the system called "*danka*" that obliged every Japanese family to register with the local Buddhist temple [81]. The popular story of the "Forty-seven *rōnin*" that took place in 1702 tells us how the government handled with a heavy hand every attempt to act outside the laws: just note how all the "heroes" of the story were forced to commit ritual suicide.

Today, we would define the *bafuku* as a harsh dictatorship: it was ruthless against everything it perceived as a threat. Among other things, it forbade Christianity, believed to be a tool of foreigners to gain a foothold in Japan and, eventually, dominate it. But, mainly, the *bafuku* was engaged in playing the game that the Japanese describe with the saying, "The nail that sticks out gets hammered down." It intervened to make sure that no competing force, warlords, foreigners, or commercial companies would become strong enough to threaten the central power.

A dictatorship, sure, but it must be said that the *bafuku* ensured an environment where commerce and craftsmanship could develop and flourish. Agriculture could provide food for the whole population and the Japanese developed a lively economy based on commerce along the "five routes" (*Gokaidō*) that linked the capital, Edo, with the main cities of the islands. And Japan was not just a land of warriors and peasant, there were people whom we could identify as our concept of "middle class," merchants, artists, craftsmen, and literates. They lived in a simple world, dressed in simple cotton kimonos, their only drink was *sake*, and wherever they wanted to go, they had to walk there on their own feet. But they seemed to be able to live a fulfilling life. They enjoyed nature, poetry, literature, music, and each other's company, just think of the poetry of Matsuo Bashō (1644–1694), still known all over the world. A good visual impression of that period is the delicate and beautiful movie *Miss Hokusai* (2016).

In terms of managing the ecosystem, Japan was forced to develop a self-contained economy that produced what the system needed with minimal or no imports from abroad, what we call today a "circular economy." [82]. It was obtained mainly by a bottom-up approach where the government does not seem to have directly intervened. Gerald Marten describes how the Japanese rose to the challenge of deforestation during the Edo period [83]:

> Japan responded to this environmental challenge with a "positive tip" from unsustainable to sustainable forest use that began around 1670…. The central role of catalytic actions and mutually reinforcing positive feedback loops, local community, outside stimulation and facilitation, letting nature and natural social processes do the work, demonstration effects, social/ecological coadaptation, and using social/ecological diversity and memory as resources. It is difficult to single out the initial tipping point with certainty, but it seems to have derived from the centuries-old tradition of cooperation among villagers for protection against bandits, allotting rice fields and irrigation water, and storing rice.

These traditions of collaboration and agreement affected all the sectors of the Japanese economy. It is fascinating to read about the details of how everything was reused and recycled: candles, clothing, cooking pots, tools, brooms, umbrellas, and much more [84]. Note also that since the government had to renounce to the temptations of military adventures abroad, it had no need of cannon fodder and no reason to push the population to grow. No active top-down birth control policies seem to have been ever enacted, but the Japanese population seemed to be able to use mainly natality control to keep population stable, although in some cases it was necessary to recur to abortion or infanticide [78].

You can see here a clear example of how a complex system reacts to external perturbations by using its internal feedbacks. The system could attain sustainability just because it was complex and it had the resources and the mechanisms to adapt. Probably, it would not have worked as well—perhaps not at all—if it had been imposed by the government from above. And note that the Japanese peasants were doing exactly what their European counterparts had been doing to manage their commons: a tangle of rules, customs, cultural practices, and collective goodwill generated a situation in which nobody could overexploit the commons in the sense that Hardin had described. It was not because of legal punishments, nor because of fences: it was because nobody could afford to place him or herself alone against the whole community.

All this is not meant to provide a blueprint for what we should do in the future. The Edo culture was characteristic of a specific period and of a specific area and, obviously, we would never be able to recreate Edo Japan in the modern world, even if we were convinced that it was worth doing that. Discussing that age is, mainly, a demonstration of feasibility. The Edo experience shows that it was possible to create a society that thrived for two centuries or more in conditions of zero growth and sustainability. It was, under several respects, a brutal dictatorship, but it was also a sophisticated and refined culture that attained, among other things, levels of literacy that were superior to those of the European society of the time. Note how the system was finely structured and optimized: it was not purely bottom-up nor purely top-down. The government ensured stability by a top-down management, the people ensured flexibility by a bottom-up management. No need of a big brother to micro-manage the commons, nor it was a free-for-all libertarian paradise. It was a machine that had attained the "self-organized criticality" conditions that we discussed in an earlier chapter of this book.

If Japan could attain economic stability, it means that it is possible to do that in other conditions, in different cultures, maybe even at the worldwide level. What the story of Edo Japan tells us is in line with what we know about complex systems: they tend toward stability. In other words, our current fixation on growth may be just a quirk of history, destined to fade away in the future as we find ourselves forced to live within the limits of the Earth's ecosystem. But there one condition that we badly need for that: it is *peace*, as the Edo experience tells us.

Surely, reaching such a condition will take time and efforts and, at present, we have little or no idea of what kind of political system could manage the planetary commons for the good of all humankind. Most likely, we'll have to go through some kind of "Seneca bottleneck" before we learn how to do that, but it is not impossible to attain sustainability, especially because it is unavoidable.

Returning After Collapse: The Seneca Rebound

Fig. 4.8 This is the way we tend to see Europe during the "Dark Ages"—a depopulated land of isolated castles. The image shows the Hermitage Castle in Liddesdale, Scotland, in a print made in 1814 (https://www.flickr.com/photos/126409951@N04/14772362853. It is the prototypical sinister castle, probably haunted by the appropriate ghosts (in this case, said to be Mary, Queen of the Scots)

Imagine Europe at the start of the period we call the "Dark Ages," more correctly "Late Antiquity." In 650 AD, the European population has shrunk to some 18 million people [85], less than half of what it had been during the high times of the Roman Empire and enormously smaller than it is today, some 700 million people. The Europe of that age was a forested region, nearly empty of people, where nothing especially interesting happened except for the squabbles of local warlords fighting each other. No one at that time could have imagined that, in less than a millennium, the descendants of the inhabitants of that backward peninsula of the Eurasian continent would start the bold attempt of conquering the world and, eventually, succeed at that. By end of the 19th century, practically all the world was under the direct or indirect control of European countries or of their American offspring, the United States. Under some respects, the situation has not changed much today.

The conventional explanation for the European success at conquering the world has to do with the "white man's burden", a term invented by Rudyard

Kipling in 1899. According to this idea, the European domination was a sort of "manifest destiny" generated by the superior genetic or cultural qualities of the European people who turned out to be smarter, more laborious, better organized, and generally more efficient than the populations of the rest of the world, supposed to be lazy, disorganized, uncultured, and in the grip of superstitions.

It is surely flattering for Europeans to think that they are smarter than everybody else, but it is also an interpretation that is not supported by data: Richard Dawkins actually argues for the opposite in his book *Guns, Germs, and Steel* (1997). Indeed, when non-Europeans were given a chance to confront the Europeans using the same weapons, the European superiority was far from being assured. Some historical cases include the battle of Adwa in 1895 when Ethiopian forces destroyed an Italian invading contingent, and the battle of Tsushima, in 1905, when a Japanese fleet defeated a Russian fleet during the Russo-Japanese war of 1904–1905. In more recent times, we have the example of Vietnam, where the mighty United States had to admit defeat to the Vietnamese forces in 1975.

But these were exceptions to the general rule that sees Europeans dominate almost everywhere in the world and the list of the battles and of the wars won by European or American forces against non-European ones would probably require several pages. So, what led Europeans to have so much success? Without pretending to have the definitive explanation, I think I can propose that it is not a question of genetic or cultural factors but rather that it was caused by a phenomenon that I call the "Seneca Rebound"—the fact that a society, a state, or an organization can restart growing after collapse at a faster speed than before the collapse. In this case, Europe may have obtained a decisive advantage in a specific historical period because of a combination of geographical and historical factors that caused its population to "rebound" at the right moment. It happened when the technologies needed to expand all over the world had been developed and could be used for that purpose.

A rebound is something that comes after collapse and there is no doubt that Europe has known economic and population collapses over its long history. There is evidence of an early European collapse that took place during the Neolithic, in the 5th millennium BCE [86]. Then, of course, there was the collapse of the Western Roman Empire that started around the 3rd century CE. Moving onward in time, we have the terrible collapse of the mid 14th century, when famines, wars, and the plague epidemics known as the "Black Death" wiped out an estimated 30% to 50% of the European population of the time [85]. There was another collapse during the mid 17th century, in correspondence to the "Little Ice Age" although less pronounced and less destructive than the others.

So, we have a total of four major collapses over European history and each collapse, so far, was followed an economic rebound and by a rapid population growth. There are no quantitative data for the first two rebounds, but a visual impression for the events that took place during the past millennium can be seen in a paper by William Langer, published in 1964 [87] (Fig. 4.9).

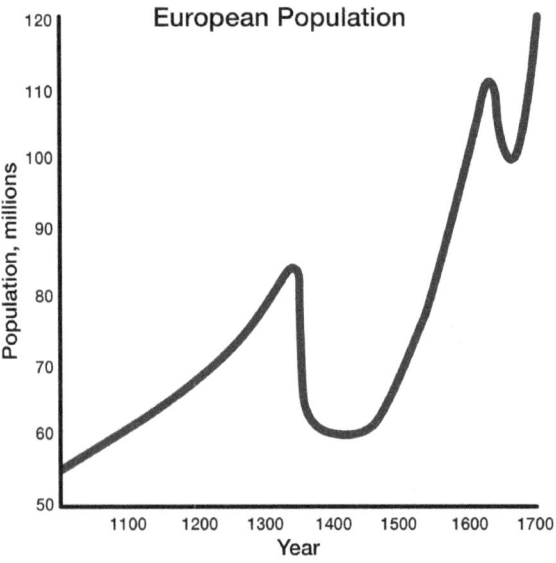

Fig. 4.9 Graph from William E. Langer, 1964 [87]. Note how growth is faster after the collapse. This is what I call the "Seneca Rebound"

These are the data: how do we explain them? The first question usually asked is what caused the collapses, but it may be an ill-posed one. It is typical of complex systems to behave in a complex manner and that may generate a series of feedback effect that may mistakenly be taken as the "cause" of the collapse. For instance, the Neolithic collapse of Europe was accompanied by an invasion of nomads (the "Yamnaya") [88] and we all know how the Roman Empire saw its territory swept by wave after wave of barbarian populations during the last phases of its existence. In both cases, the invasions have been proposed as the cause of the collapse, but note that no such invasions took place in correspondence with the two later European crashes, so we are justified to think that the previous invasions were opportunistic reactions to an already weakened society.

Then, consider climate change: it is a typical cause reported for civilization collapses, but its effects have been ruled out for the Neolithic collapse [86] and no significant temperature changes are reported in correspondence to the decline and collapse of the Western Roman Empire [89]. Instead, in the case of the two more recent collapses in Europe, there is evidence of cold spells that damaged agriculture, possibly generated by volcanic eruptions. So, so maybe climate change caused these collapses? It is possible but, as usual, complex systems defy simple interpretations in terms of cause and effect. Maybe the population decline was generated by atmospheric cooling, but it may also be that the population drop cooled the climate as the result of reforestation—another case of reinforcing feedback in a complex system. Indeed the data show a small decline in atmospheric CO_2 concentration in the centuries after the Black Death in Europe [89]: it may have contributed to the cooling. The effect is stronger and clearer for the great crash in the populations of the New World [90], occurring in a later period. Overall, it seems that the European collapses are mostly the result of internally generated feedbacks in societies that were growing so fast that they had outpaced the capability of the resources they were exploiting to keep pace.

In any case, the point is not so much what caused the collapses but the remarkably rapid recovery that followed them: what I call here the "Seneca Rebound." The reasons for the rebound are reasonably clear: depopulation frees resources that can be exploited for a new phase of rapid growth. Before the fossil fuel age, societies had two main natural resources to exploit: fertile soil and forests. Both tend to be overexploited: forests are cut faster than trees can regrow and the fertile soil is eroded and washed to the sea faster than it can reform. That generates a decline of agriculture and the result is not just an end to population growth, it is a ruinous collapse resulting from famines and epidemics. The loss of revenues from forests weakens the state and the result is internecine wars which also hasten the collapse. But the disappearance of a large fraction of the population frees cultivated land for forests to regrow and that regenerates the soil. Then, when the population starts regrowing, people find in the new forests a near-pristine source of wood and, once cut, of fertile soil. Trees provide the wood for ships and the charcoal made from wood provides the material needed to make steel for weapons. The cycle restarts and it may go faster than the earlier one because society still remembers the social structures and the technologies of the previous cycle.

The cycles of deforestation and reforestation are evident in Europe: both the Roman Empire and the Medieval society had badly overexploited their forests and the reforestation after the collapses freed resources that could be used for the population to grow and expand beyond the earlier borders. The phenomenon was not unique to Europe but, as always, success is a question of timing, opportunities, and a little luck. The Europeans found themselves rebounding forward in a moment when they had the right technologies to expand worldwide and while the other, potentially competing, civilizations were unable to stop them.

On the opposite side of the Mediterranean, the Arab civilization was socially and technologically as sophisticated as the European one, but its climate did not allow forests to grow fast enough to generate the same rebound seen in Europe. The American civilizations we call "pre-Columbian" had forests, but they hadn't yet developed the technologies of steel and of oceanic ships—they also lacked horses for transportation and as a military weapon. The Chinese, instead, had the technologies and also the forests and they could have wrestled with Europe for the control of the world. During the 12th-13th centuries, an outbreak of the same plague that affected Europe caused a decline in the Chinese population that was followed by the Mongol invasion. Then, the Chinese economy experienced a rebound: the population restarted growing and the age of "treasure voyages" started in the early 1400 s, during the Ming dynasty, with fleets of ships exploring the lands around China. But the Chinese exploration phase soon stopped when the central government forbade all oceanic travels. We can only wonder what would have happened if the Chinese government had continued to support overseas exploration. Maybe Columbus would have found Chinese-speaking people when he landed in the New World. But that is the way history works.

During the Middle Ages, Europe didn't have a central government, as China did, so there were no brakes applied to the military expansion of the European states, competing with each other to conquer new lands. The first phase of European expansion came with the Crusades—the first one took place in 1095. But the real push forward was with the rebound after the Black Death of the mid-1300s: it was called the "age of explorations" and we know how the Europeans managed to expand over most of the Americas and in Africa. After the latest collapse, the one that took place in the mid-17th century [85] there was another burst of economic growth which ushered in the age of coal and, with it, the period defined as "The Age of Divergence" by Kenneth Pomeranz with the book he published in 2000 [91], when Europe truly became the dominating world power. Right now, Europe is declining again, maybe there will be a new phase of collapse and rebound in the future.

These considerations are qualitative, but it is possible to see the Seneca Rebound as an engine that propels civilizations forward in bursts. If this is the case, can we expect a rebound if the world's civilization goes through a new Seneca Collapse in the coming decades? If previous history can serve us as a guide, it might happen. Of course, it is possible that the upcoming collapse will be so bad that humankind will never return to the complexity of the civilization it managed to create during the 20th and 21st century. For all we know, the effects of the destruction we are wreaking on the ecosystem could cause humans to go extinct, the ultimate Seneca Collapse. But a much more interesting case, and I would also say a more probable one, is that the coming collapse will be just one more of the series of previous collapses that affected human civilizations: it might lead to a new rebound. Would that be really possible in a world badly depleted in terms of mineral resources and subjected to extensive ecosystem damage?

As we saw in earlier chapters of this book, a complex system is an entity that lives on an energy flow. A civilization needs energy to survive and, the more energy it can get, the more complex and structured it can be. The problem we are examining here is whether a sufficient energy flow of energy can be maintained for civilization to keep at least some of the characteristics it has today, for instance the electronic treatment and storage of information, a worldwide Internet, automation, scientific research, and more.

Today, our civilization is maintained by a flow of some 18 TW of primary energy, mainly (ca. 85%) produced by the combustion of fossil fuels [92]. The rest is provided in part by nuclear fission (ca. 6%) and by a mix of renewable technologies such as hydroelectric, photovoltaic, wind, and others. A civilization of complexity comparable to ours cannot exist without access to a comparable flow of energy. The resources that powered ancient civilizations, wood and animal power, created remarkably sophisticated societies, but none endowed with the technological level we have reached. So, the first question is what would happen to the current energy sources in case of a collapse of the world's economic system.

We can be reasonably certain that fossil fuels won't survive the Seneca bottleneck. The deposits of these fuels have been badly depleted over a couple of century of exploitation and, today, it is possible to maintain production only by means of extremely sophisticated technologies and large inputs of financial and human capital. An extensive economic and social crisis, coupled with wars and civil unrest, could easily send the fossil fuel industry down a death spiral from which it might never re-emerge. It would be the end of the "fossil age," at least until the Earth manages to re-create them, but that would take millions of years.

The situation is even more difficult for nuclear energy. First, nuclear energy is also affected by depletion in the same way as fossil energy. The high-concentration mineral resources of uranium have been largely consumed by the exploitation of the 20th century and a future civilization attempting to restart with fission reactors would have to reckon with the lack of inexpensive uranium resources. Perhaps they could use our abandoned nuclear warheads, but it is an iffy proposition, to say the best. They might try to jump start to the much more expensive and complex technology of "fast" reactors, able to breed fissile material from non-fissile isotopes but this is, again, a difficult proposition, especially if starting from scratch. A further, and perhaps worse, problem for nuclear energy is that an abandoned nuclear plant is at serious risk of going into meltdown if it loses active cooling. Typically, the fuel will melt because of its residual radioactivity and then the reactor vessel may build up enough pressure for it to explode, spreading radioactive material all over. This is what happened to one of the reactors of Fukushima, hit by the tsunami of 2011. In the case of an extended breakdown of the societal structure, the current reactors—there are about 500 of them—are all at risk of meltdown, a collective disaster with nearly unimaginable consequences. Even if that can be avoided, nuclear reactors remain vulnerable to military action, terrorism, or sabotage [93, 94]. In case of a major economical collapse, with the associated social and strategic unrest, nuclear reactors could become a major burden rather than an asset and those destroyed by meltdown would remain radioactive traps for centuries—hardly something that would encourage our descendants to restart with the technology.

Things look much better if we examine the third leg of the current energy supply: renewables. On all counts, renewables are more resilient than both nuclear and fossil technologies. Renewables are not subjected to fuel depletion, even though, of course, the plants wear out and need to be periodically replaced. But most of the materials used in a renewable plant can be recycled and these technologies need little or no rare minerals. Photovoltaic (PV) panels use only silicon and aluminum, both very abundant on the Earth's crust, plus traces of other common minerals—that includes some silver, but it is not essential to their functioning [95]. Wind plants use rare earths for their magnets but, also in this case, alternatives are available and it is also possible to recycle the materials of an old plant to build a new one. Renewable plants are also long lasting. One of the first PV plants in Italy was installed in 1984 and, more than 30 years later, in 2016, it was still working, having lost just about 10% of its initial efficiency [96]. Of course, the electronic parts of a PV plant need to be replaced at shorter intervals, but even without an inverter the panels can still provide DC power: it is what is needed, for instance, to recharge batteries.

In general, PV plants can take a lot of damage and continue functioning. I personally witnessed how a plant in Italy was hit by a twister that turned it into something that looked like a Mad Max movie scene of broken panels scattered all over. But when the sun shone again, the remaining panels, although damaged, still produced more than 50% of the power that the plant had been producing before the disaster. The plant could be rapidly repaired and now it works at full power. The situation may be more difficult for the modern generation of wind plants: tall wind towers can fall in conditions of exceptionally strong winds and, in that case, there is little that can be done except rebuild the plant from scratch. Instead, hydroelectric plants can last a long time and are very resilient to damage.

Overall, it is possible that the renewable infrastructure of a country may survive a crisis that could include major military operations, civil disturbance, and ecosystem collapses. Our descendants could re-emerge on the other side of the Seneca bottleneck relying on these plants to produce electric power. This power could be used to build new plants to replace the old ones as they wear out. The diffuse legend that renewable energy needs fossil energy in order to keep going is just that: a legend [97]. Over the course of their life, renewable plants produce much more energy than it is needed to create their replacement. So, it would be possible for our descendants to have a good supply of electric power using the renewable technologies that our society has developed.

That leaves open the question of mineral resources: a future civilization would not have the cheap ores that ours has depleted. Yet, our descendants would have large amounts of minerals already extracted that they could salvage from the ruins of our civilization. It is nothing new: during the Middle Ages, people would scavenge Roman ruins for stone and metals. From our waste, our descendants could have plenty of metals of all kinds and their probably smaller population wouldn't need so much of them as we do nowadays. That would be sufficient to jump start a new civilization.

Of course, our ruins could not last forever as sources of minerals: just as we are not mining Roman ruins anymore, our descendants would need to find new sources. Since they won't have the same high-grade ores we had, they would be constrained in terms of the mineral resources they could use, but they would still have good strategies to keep going. As I discuss in my book *Extracted*, [98] the Earth's crust contains abundant silicon for electronic devices and for photovoltaic panels, plenty of metals such as iron, titanium, aluminum, and magnesium for structural applications and, of course, plenty of silicon oxide for glass and the like. As conductors, copper, too rare, would have to be replaced with aluminum. Other technologies should have to be re-designed to use none or very little of the rare metals we use nowadays, from gallium for semiconductors to rare earths for magnetic materials. It would be a long term challenge that, nevertheless could be met, at least in principle. There is no need for humankind to return to subsistence agriculture or to hunting and gathering although, of course, it might be argued that it will happen and even that it would be a good idea.

There is another possibility worth discussing: could humankind mine space bodies to replace the dwindling ores on out planet? This would be enormous expensive and in many cases it would be useless even if we could afford to pay the cost. The concentrations of elements we call "ores" is a characteristic of geologically active bodies and we know on only one such bodies: our Earth. There are no ores on the Moon or on the asteroids—maybe on Mars, but we have no evidence of that, so far. So, mining space bodies to bring minerals to Earth makes little sense. Nevertheless, there may be a logic in the idea if we change the target market from the Earth to space. Asteroids are rich in elements such as iron, nickel, aluminum, titanium, silicon and even carbon and water in the form of ice. These minerals are not there in the form of ores, but they form a sufficiently large fraction of some asteroids that extracting and purifying them could be possible. Take also into account that space is rich in solar energy that can be transformed into electric power by PV panels and that in space you have little to worry about pollution. Of course, putting together a mining industry in space is a task which was never attempted so far and the unknowns are enormous. One thing is clear: it is not a task for humans. Humans cannot live in space unless they bring with them expensive and complex equipment and it is extremely difficult to shield them from dangerous high-energy radiation [99]. Instead, space is a good place for robots which can do the same things human can do in a better and cheaper way. And these robots could be made, at least in part, from materials obtained from asteroids. Our robot-children have a chance to inherit the solar system and they could build a completely new, silicon-based, civilization [92].

The future is beautiful because it is always full of possibilities and what we do now will echo in eternity. As Seneca said in one of his letters,

"Every new beginning comes from some other beginning's end."

References

1. Fleischmann, M., Pons, S.: Electrochemically induced nuclear fusion of deuterium. J. Electroanal. Chem. Interfacial Electrochem. **261**, 301–308 (1989)
2. Bardi, U.: Cold fusion and I. *Cassandra's Legacy* (2012). https://cassandralegacy. blogspot.com/2012/11/cold-fusion-and-i.html. Accessed 29 January 2019
3. Berlinguette, C.P., et al.: Revisiting the cold case of cold fusion. Nature **570**, 45–51 (2019)

4. What is ITER? ITER (2018). https://www.iter.org/proj/inafewlines#6. Accessed 31 January 2019
5. Cohen, H.: "Table of temperatures, power densities, luminosities by radius in the Sun". Contemporary Physics Education Project (1998). http://webarchive.loc.gov/all/20011129122524/http%3A//fusedweb.llnl.gov/cpep/chart_pages/5.plasmas/sunlayers%2Ehtml. Last accessed Aug 30, 2019
6. Bardi, U.: The Oil Drum: Europe|The Universal Mining Machine. The Oil Drum (2008). http://www.theoildrum.com/node/3451. Accessed 24 August 2013
7. O'Neill, G.K.: The colonization of space. Phys. Today **27**, 32–40 (1974)
8. Dyson, F.J.: Search for artificial stellar sources of infrared radiation. Science **131**, 1667–1668 (1960)
9. Newman, W.I., Sagan, C.: Galactic civilizations: population dynamics and interstellar diffusion. Icarus **46**, 293–327 (1981)
10. Meadows, D. H., Meadows, D. L., Randers, J. & Bherens III, W. The Limits to Growth. (Universe Books, 1972)
11. Heinlein, R.A.: Where to? Galaxy **3**, 13–22 (1952)
12. Solow, R.: Technical change and the aggregate production function. Q. J. Econ. **70**, 65–94 (1956)
13. Nordhaus, W.: Lethal models. Brook. Pap. Econ. Act. **2**, 1–59 (1992)
14. Warr, B.S., Ayres, R.U.: Evidence of causality between the quantity and quality of energy consumption and economic growth. Energy **35**, 1688–1693 (2010)
15. Daly, H.E.: Steady-State Economics: The Economics of Biophysical Equilibrium and Moral Growth. W.H. Freeman (1977)
16. Giarini, O., Laubergé, H.: The Diminishing Returns of Technology. Pergamon Press, Oxford (1978)
17. Tainter, J.A.: Social complexity and sustainability. Ecol. Complex. **3**, 91–103 (2006)
18. Niccolai, J.: Intel pushes 10 nm chip-making process to 2017, slowing Moore's Law|InfoWorld. InfoWorld (2015). https://www.infoworld.com/article/2949153/hardware/intel-pushes-10nm-chipmaking-process-to-2017-slowing-moores-law.html. Accessed 30 January 2019
19. Giampietro, M., Mayumi, K.: The Biofuel Delusion: The Fallacy of Large Scale Agro-Biofuels Production. Earthscan, London (2009)
20. Bardi, U.: The biodiesel disaster: why bad ideas are always so successful? Cassandra's Legacy (2019). https://cassandralegacy.blogspot.com/2019/02/the-biodiesel-disaster-why-bad-ideas.html. Accessed 8 February 2019
21. Jevons, W.S.: The Coal Question, 2nd revised edn. Macmillan and Co, London (1866)
22. Saunders, H.D.: A view from the macro side: rebound, backfire, and Khazzoom–Brookes. Energy Policy 439–449 (2000)
23. Bardi, U., Perissi, I., Csala, D., Sgouridis, S.: The Sower's way: a strategy to attain the energy transition. Int. J. Heat Technol. **34** (2016)

24. Ettinger, R.C.W.: Man Into Superman: The Startling Potential of Human Evolution–and How to Be Part of It. Ria University Press, Palo Alto (2005)
25. Kurzweil, R.: The Singularity Is Near: When Humans Transcend Biology. Viking (2005)
26. Dyson, G.: Childhood's end. Edge (2019)
27. Drone Warfare: The Bureau of Investigative Journalism (2019). https://www.thebureauinvestigates.com/projects/drone-war. Accessed 10 March 2019
28. Russel, S.: Slaughterbots (2017). https://www.youtube.com/watch?v=9CO6M2HsoIA. Last Accessed Aug 30, 2019
29. Haworth, A.R., Sagan, S.D., Valentino, B.A.: What do Americans really think about conflict with nuclear North Korea? The answer is both reassuring and disturbing. Bull. At. Sci. 1–8. https://doi.org/10.1080/00963402.2019.1629576 (2019)
30. Tillerson: Sanctions 'really starting to hurt' North Korea. Time (2018). http://time.com/5107417/tillerson-sanctions-north-korea/. Accessed 19 January 2019
31. BLU-114/B "Soft-Bomb" Federation of American Scientists (2000). https://web.archive.org/web/19991011170902/http://www.fas.org/man/dod-101/sys/dumb/blu-114.htm. Accessed 25 March 2019
32. Ristuccia, C.A.: 1935 sanctions against Italy: would coal and crude oil have made a difference (1997), University of Oxford Working Paper. https://www.economics.ox.ac.uk/oxford-economic-and-social-history-working-papers/1935-sanctions-against-italy-would-coal-and-crude-oil-have-made-a-difference. Last Accessed Aug. 30, 2019
33. Bardini, C.: Senza carbone nell'età del vapore. Gli inizi dell'industrializzazione italiana. Bruno Mondadori (1998)
34. Kopp, C.: Considerations on deception techniques used in political and product marketing. Aust. Inf. Warf. Secur. Conf. (2006). https://doi.org/10.4225/75/57a80fdfaa0ccm Last Accessed March 19, 2019
35. Suskind, R.: Faith, Certainty and the Presidency of George W. Bush. The New York Times (2004). https://www.nytimes.com/2004/10/17/magazine/faith-certainty-and-the-presidency-of-george-w-bush.html. Last Accessed Aug 30, 2019
36. Bardi, U.: When fake news kill. Mata Hari, the spy who never was. Chimeras. https://cassandralegacy.blogspot.com/2017/07/when-fake-news-kill-mata-hari-never-was.html. Accessed 19 June 2019
37. Docherty, G., Macgregor, J.: Prolonging the agony: how the Anglo-American establishment deliberately extended WWI by three-and-a-half years, Trine Day, 2018, SBN-10: 1634241568
38. Wikisource, the free online library https://en.wikisource.org/wiki/Moscow_Declarations. Accessed 5 Mar 2019
39. Beschloss, M.R.: The Conquerors: Roosevelt, Truman, and the Destruction of Hitler's Germany, 1941–1945. Simon & Schuster (2002)

40. Unz, R.: American Pravda: our deadly world of post-war politics. Unz Rev. (2018). http://www.unz.com/runz/american-pravda-our-deadly-world-of-post-war-politics/#comments. Accessed 26 June 2019
41. Haynes, J. E. & Klehr, H. Venona: decoding Soviet espionage in America, Yale Nota Bene (2000)
42. Marsh, C.: Social harmony paradigms and natural selection: Darwin, Kropotkin, and the metatheory of mutual aid. J. Public Relat. Res. **25**, 426–441 (2013)
43. Akerstrom, M.: Betrayal and Betrayers. Routledge (2017). https://doi.org/10.4324/9781351316804
44. Myerson, R.B.: Game Theory: Analysis of Conflict. Harvard University Press (1997)
45. Rapoport, A., Chammah, A.M.: Prisoner's dilemma: a study in conflict and cooperation. University of Michigan Press (1970)
46. Poundstone, W. Prisoner's Dilemma. Doubleday (1992)
47. Bardi, U. Facing the climate bear: the Camper's dilemma. Cassandra's Legacy (2017). https://cassandralegacy.blogspot.com/2017/06/facing-climate-bear-campers-dilemma.html. Accessed: 4th March 2019
48. Vidal, J.: We need development': Maldives switches focus from climate threat to mass tourism. The Guardian (2017). https://www.theguardian.com/environment/2017/mar/03/maldives-plan-to-embrace-mass-tourism-sparks-criticism-and-outrage. Last Accessed 30 Aug, 2019
49. Bardi, U. Islands not sinking: climate change demonstrated to be a hoax. Cassandra's Legacy Blog (2018). https://cassandralegacy.blogspot.com/2018/02/islands-not-sinking-climate-change.html. Accessed 8 June 2019
50. Nerem, R.S., et al.: Climate-change-driven accelerated sea-level rise detected in the altimeter era. Proc. Natl. Acad. Sci. USA **115**, 2022–2025 (2018)
51. Kench, P.S., McLean, R.F., Nichol, S.L.: New model of reef-island evolution: Maldives. Indian Ocean. Geol. **33**, 145 (2005)
52. As climate change threatens islands, Kiribati's president plans development. CBS News (2017). https://www.cbsnews.com/news/climate-change-kiribati-president-taneti-maamau/. Last Accessed Aug 30, 2019
53. Transcript: The Vice-Presidential debate. The New York Times (2008). https://www.nytimes.com/elections/2008/president/debates/transcripts/vice-presidential-debate.html. Last Accessed Aug 30, 2019
54. Jordan, K.N., Sterling, J., Pennebaker, J.W., Boyd, R.L.: Examining long-term trends in politics and culture through language of political leaders and cultural institutions. Proc. Natl. Acad. Sci. USA **116**, 3476–3481 (2019)
55. Chen, S.: Is China's plan to use a nuclear bomb detonator to release shale gas in earthquake-prone Sichuan crazy or brilliant? South China Morning Post (2019). https://www.scmp.com/news/china/science/article/2183466/chinas-plan-use-nuclear-bomb-detonator-release-shale-gas. Accessed 7 April 2019

56. Berman, A.: Alternative facts about OPEC & U.S. shale from the wall street journal - Art Berman. Art Berman (2018). https://www.artberman.com/alternative-facts-about-opec-u-s-shale-from-the-wall-street-journal/. Accessed 20 February 2019

57. Hotelling, H.: The economics of exhaustible resources. J. Polit. Econ. **39**, 137–175 (1931)

58. Goeller, H., Weinberg, A.: Age of substitutability: what do we do when the mercury runs out. Am. Econ. Rev. **68**, 1–11 (1978)

59. Bison in the Journals: Discovering Lewis & Clark. http://www.lewis-clark.org/article/443#jump. Accessed 21 January 2019

60. Doughty, C. E., Wolf, A., Field, C.B.: Biophysical feedbacks between the Pleistocene megafauna extinction and climate: the first human-induced global warming? Geophys. Res. Lett. **37** (2010)

61. Scott Baker, C., Clapham, P.J.: Modelling the past and future of whales and whaling. Trends Ecol. Evol. **19**, 365–371 (2004)

62. Demetrion, D.: Japan's squid industry in crisis amid record low catches. The Telegraph (2019). https://www.telegraph.co.uk/news/2019/01/21/japans-squid-industry-crisis-amid-record-low-catches/. Last accessed Aug 30, 2019

63. Bardi, U.: Perché il colibrì è l'animale più pericoloso che esista. Effetto Cassandra (2018). https://ugobardi.blogspot.com/2018/02/perche-il-colibri-e-lanimale-piu.html. Accessed 1 February 2019

64. Comby, J.: La Question climatique. Genèse et dépolitisation d'un problème public. http://journals.openedition.org/sociologie. Raison D'Agir (2015)

65. Bardi, U.: Chemistry of an Empire: the last Roman empress. Cassandra's Legacy (2011). https://cassandralegacy.blogspot.com/2011/12/chemistry-of-empire-last-roman-empress.html. Accessed 16 Nov 2018

66. Oost, S.I.: Galla Placidia Augusta. A Biographical Essay. The University of Chicago Press (1969). https://doi.org/10.1017/s0009840x00262501

67. Hardin, G.: The tragedy of the commons. Science (80–). **162**, 1243–1248 (1968)

68. Ostrom, E.: Governing the commons: the evolution of institutions for collective action. Cambridge University Press, Cambridge (1990)

69. Perissi, I., Lavacchi, A., Bardi, U., El Asmar, T.: Dynamic patterns of overexploitation in fisheries. Ecological Modeling, **359**, 10 Sep 2017, 285–292

70. Volterra, V.: Fluctuations in the abundance of a species considered mathematically. Nature **118**, 558–560 (1926)

71. Forrester, J.W.: Counterintuitive behavior of social systems. Simulation **16**, 61–76 (1971)

72. Forrester, J.: World Dynamics. Wright-Allen Press (1971)

73. Meadows, D.H.: Leverage points: places to intervene in a system. donellameadows.org (1999). http://leadership-for-change.southernafricatrust.org/downloads/session_2_module_2/Leverage-Points-Places-to-Intervene-in-a-System.pdf. Last Accessed, Mar 19, 2019

74. Confino, J.: It is profitable to let the world go to hell. The Guardian (2015). https://www.theguardian.com/sustainable-business/2015/jan/19/davos-climate-action-democracy-failure-jorgen-randers. Accessed 18 June 2019

75. Mattei, U.: Beni comuni. Un manifesto. Laterza (2011)

76. Moxnes, E.: Not only the tragedy of the commons: misperceptions of feedback and policies for sustainable development. Syst. Dyn. Rev. **16**, 325–348 (2000)

77. Hieber, R., Hartel, I.: Impacts of SCM order strategies evaluated by simulation-based 'Beer Game' approach: the model, concept, and initial experiences. Prod. Plan. Control **14**, 122–134 (2003)

78. Eng, R.Y., Smith, T.C.: Peasant families and population control in eighteenth-century Japan. J. Interdiscip. Hist. **6**, 417 (1976)

79. Bardi, U.: The Population problem: should the Pope tell people to stop breeding like rabbits? Cassandra's Legacy (2016). https://cassandralegacy.blogspot.com/2016/04/the-population-problem-should-pope-tell.html. Accessed 5 March 2019

80. Oishi, S.: Edo Jidai (The Edo Period). Chuko Shinso **476** (1977). http://www.grips.ac.jp/teacher/oono/hp/lecture_J/lec02.htm. Last Accessed Aug 30, 2019

81. Tamamuro, F.: The development of the temple-parishioner system. Jpn. J. Relig. Stud. **36**, 11–26 (2009)

82. Hosomi, M.: The edo period created the sound material-cycle society. Clean Technol. Environ. Policy **17**, 2091 (2015)

83. Marten, G.: Environmental tipping points: a new paradigm for restoring ecological security. J. Policy Stud. **20**, 75–87 (2005)

84. Greve, G.: Recycling and reuse. Edo - The Edopedia (2014).https://edoflourishing.blogspot.com/2013/11/business-in-edo.html. Accessed 22 June 2019

85. Durand, J.D.: Historical estimates of world population: an evaluation. Popul. Dev. Rev. **3**, 253 (1977)

86. Shennan, S., et al.: Regional population collapse followed initial agriculture booms in mid-Holocene Europe. Nature Communications **4**, Article number: 2486 (2013)

87. Langer, W.L.: The black death. Sci. Am. **210**, 114–121 (1964)

88. Kristiansen, K., et al.: Re-theorising mobility and the formation of culture and language among the Corded ware culture in Europe. Antiquity **91**, 334–347 (2017)

89. Buntgen, U., et al.: 2500 years of European Climate variability and human susceptibility. Science (80-.). **331**, 578–582 (2011)

90. Koch, A., Brierley, C., Maslin, M.M., Lewis, S.L.: Earth system impacts of the European arrival and great dying in the Americas after 1492. Quat. Sci. Rev. **207**, 13–36 (2019)

91. Pomeranz, K.: The Great Divergence: China, Europe, and the Making of the Modern World Economy, Princeton University Press (2000)

92. Bardi, U.: What future for the Anthropocene? A biophysical interpretation. Biophys. Econ. Resour. Qual. **1**, 2 (2016). https://doi.org/10.1007/s41247-016-0002-z

93. Lovins, A.B., Lovins, H.L.: Brittle Power. Brick House Publishing Company (1982)

94. Vittitoe, C.N.: Did high altitude EMP cause the Hawaiian streetlight incident? Syst. Des. Assess. Notes (1989). https://www.osti.gov/biblio/6151435. Last Accessed Aug 30, 2019

95. Solé, J., García-Olivares, A., Turiel, A., Ballabrera-Poy, J.: Renewable transitions and the net energy from oil liquids: a scenarios study. Renew. Energy **116**, 258–271 (2018)

96. Bardi, U.: But what's the real energy return of photovoltaic energy? Cassandra's Legacy **47**, 133–141 (2016), Last Accessed 1st Sep 2019

97. Raugei, M., et al.: Energy return on energy Invested (ERoEI) for photovoltaic solar systems in regions of moderate insolation: a comprehensive response. Energy Policy **102** (2017)

98. Bardi, U.: Extracted: how the quest for mineral resources is plundering the planet. Chelsea Green (2014)

99. Lanzerotti, L.J.: High-energy solar particles and human exploration. Sp. Weather. **3** (2005)

100. Laherrère, J.: The hubbert curve: its strenghts and weaknesses. Oil Gas J. (2000), archived on http://dieoff.com/page191.htm. last accessed on Aug 29, 2019

101. Campbell, C.J., Laherrere, J.F.: The end of cheap oil. Scientific American. **278** (3) (MARCH 1998)

102. Odds of Dying - Data Details - Injury Facts. https://injuryfacts.nsc.org/all-injuries/preventable-death-overview/odds-of-dying/data-details/. Accessed 7 Mar 2019

5

Conclusion: Collapse as Seen in Ancient Philosophy

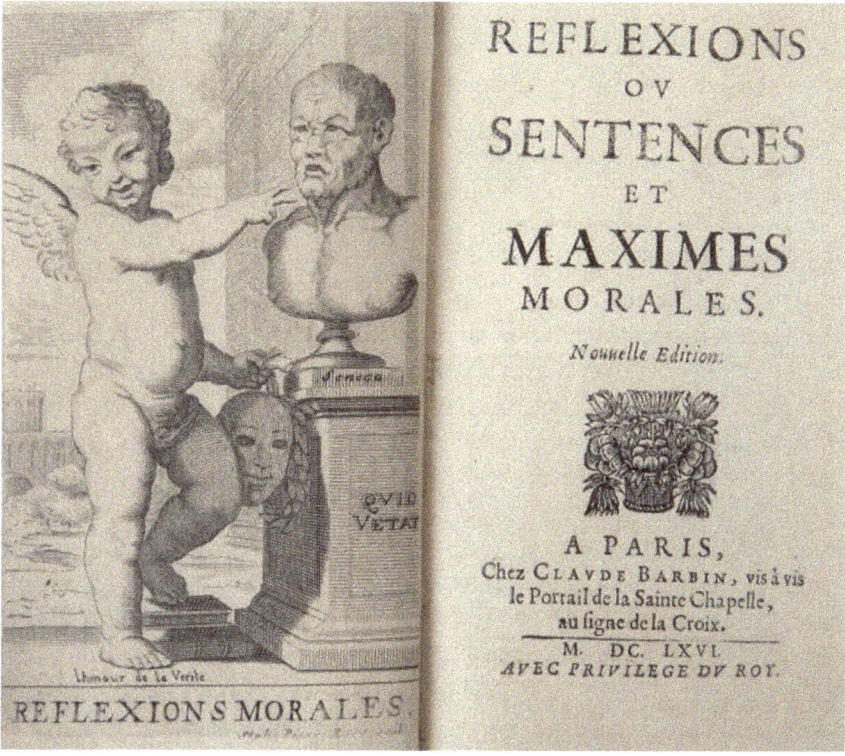

Fig. 5.1 The front cover of the Maxims of M. De La Rochefoucald, published in 1665. The illustration shows an angel and a bust that should be understood as being Lucius Seneca's face. The angel has removed the smiling mask to reveal the suffering face behind: De La Rochefoucald believed that Seneca was not sincere when he said his virtue made him impervious to pain

This book started from a sentence written some two thousand years ago by the Roman philosopher Lucius Annaeus Seneca, *"fortune is of sluggish growth, but the way to ruin is rapid."* Now that we have arrived to the end, we may return to Seneca and ask if we still learn something from him and from the school of thought he belonged to, the Stoics (Fig. 5.1).

© Springer Nature Switzerland AG 2020
U. Bardi, *Before the Collapse*,
https://doi.org/10.1007/978-3-030-29038-2_5

Just like us, Seneca and his contemporaries faced difficulties, joys, pain, happiness, successes and defeats. Just like us, they tried to find some logic in what was happening around them and they found it largely in a way of thinking that still affects us. Stoicism was a school of thought started by a Greek philosopher called Zeno, in the early 3rd century BCE. Still today, there are people who define themselves as Stoics, followers of that school and, even if it is not always recognized, Stoicism deeply affected Christianity that, eventually, replaced it as a way to see the world and as a guide for human behavior. You can also find many similarities between Stoicism and the earlier thought of Confucius and Laozi: morality, justice, sense of duty, respect for everybody, and the three principles of Daoism: Compassion, frugality, and humility.

Stoicism has a deep appeal for many of us and we can still learn something from an ancient Roman philosopher who lived in times so different from ours, and yet so similar. Seneca lived, struggled, did his best, not always behaving at the same lofty level of morality he recommended in his writings. He actively meddled in politics and he was even accused to be behind the mismanaged financial operation that caused the revolt of Queen Boudica in England in 61 CE, that almost shattered the Empire [1]. In old age, Seneca was accused of having been part of a conspiracy that aimed to dethrone Emperor Nero and replace him with Seneca himself. We'll never know if the accusation had some substance, probably it was just an excuse for a capricious emperor who wanted to get rid of his old mentor. In any case, Seneca died bravely by committing suicide in 65 CE, a story that was cited many times in history as an example of personal courage and sense of duty. Surely, he personally experienced what I call in this book the "Seneca Collapse."

In addition to Seneca as a public figure, we have Seneca the philosopher who left us a large corpus of texts. He was a learned man who loved to explore various fields of knowledge, from natural phenomena to moral rules. In this field, he never was very deep as a thinker, nor especially innovative. For us, much of what he wrote looks a little artificial, even "baroque." Yet, there is a fascination in Seneca's writing that we cannot ignore and that has kept him popular for nearly 2000 years after his death. He was steeped in the art of rhetoric, an art that in ancient times was understood as what we call today "effective communication." It consisted of arranging one's arguments in ways that were at the same time understandable and convincing. You can still see this approach in Seneca's works: he wrote as if he were speaking in court or in the *Curia*, the seat of the Roman senate. He loved to intersperse his speech with terse sentences designed with the specific objective of remaining impressed in the mind of the reader. If boldface characters had existed in

Seneca's time, I am sure he would have used them. An example is "the way to ruin is rapid," which is the refrain of this book.

Of the man, Lucius Seneca, little transpires from his writings. His inner feelings are always clouded by his rhetoric and we can clearly sense that he did not want to open his soul to the reader, as Augustine of Hippo would do in later times, perhaps the first Western author to do so. We can only vaguely imagine Seneca's feelings in the turmoil of his times, in the good and the bad moments of his personal life. Surely, he had his failings and sometimes his moral exhortations sound shrill to us. Maybe his serenity was only a mask that concealed a troubled man, as De La Rochefoucauld proposed in his *Maxims* of 1665. Or, maybe, Seneca really was able to maintain his serenity in his troubled times. His behavior at the moment of his death tells us that he was a man who could face misfortune by putting up a brave face.

As all human beings, the Stoics had their limits, but I think Seneca and others such as Epictetus and Marcus Aurelius understood a fundamental point that most of their contemporaries forgot, just as we often forget it. It is that complex systems are best dealt with by going with the flow rather than attempting to force them to take the shape we want them to. That, actually, may worsen things just like another philosopher of modern times, Jay Forrester, told us when he spoke of "pushing the levers in the wrong direction" [2]. I think we can take another sentence from Seneca's vast corpus to focus on this concept, when he says in the consolation to his mother Helvia (V) that

> No man loses anything by the frowns of Fortune unless he has been deceived by her smiles.

In extracting quotes from anyone's works we always risk oversimplifying or misrepresenting someone's thought. But, in the case of Seneca, I think that it makes sense. As I said, he loved to intersperse his texts with terse sentences summarizing his views. So, I think the sentence above can be seen as a compact rendition of the basic concept of how to deal with complex systems —what I call the "Seneca Strategy." What Seneca tells us is that you should never fall into the trap of believing that, since things have always been going in a certain way, they will keep going in that way. Your fortune is never granted and what goes up tends to go down, especially if it has been going up fast and high.

And so, the Seneca Strategy is simply about understanding that there is just so much you can do to make the world go where you would want it to go. If you are at the beginning of the asymmetric Seneca curve, the collapse has not yet come and you still have a chance to slow down the climb and avoid it altogether. If you are at the highest point, the Seneca Peak, it is too late to avoid the Seneca trap, the coming cliff, but know that the more you struggle to avoid it, the steeper you are making it. And if you are sliding down already, there is little else you can do but follow the flow.

At this point, I may cite something that another Stoic philosopher, Emperor Marcus Aurelius, said about one century after Seneca, "what we do in life, echoes in eternity," so impressive that it was repeated in a modern movie, *Gladiator*, in 2000. Falls and ascents are temporary things and, no matter how hard you are falling, there will be a rebound and that rebound will depend on what you did during the growth phase. And the rebound will depend on what you did before the fall: this is the sense of "echoing" in the future. It is part of the eternal cycles of life of the universe.

As we approach a critical time for our whole civilization, we need to think of the future of humankind rather than that of our individual existences. We can help our children and their children to have a better future by acting now in the right manner: simply by leaving to them something of the bounty that our planet has handed to us and that we have wasted so stupidly, up to now. What we do now for those who will come after us, will echo in eternity.

References

1. Bardi, U.: The queen and the philosopher: war, money, and metals in Roman Britain. Cassandra's Legacy. Available at https://cassandralegacy.blogspot.com/2018/07/the-queen-and-philosopher-hidden-story.html. Accessed 8 Mar 2019 (2018)
2. Meadows, D.H.: Leverage points: places to intervene in a system. donellameadows.org. Available at http://leadership-for-change.southernafricatrust.org/downloads/session_2_module_2/Leverage-Points-Places-to-Intervene-in-a-System.pdf (1999)

Summary: Six Things You Should Know Before Collapse

In this last part of the book, let me summarize the main points of this "Guide to collapse" in a compact reference form that I hope may be useful for you.

1. **Collapse is Not a Bug, it is a Feature (the Seneca Effect).** Some 2000 years ago, the Roman philosopher Lucius Annaeus Seneca noted that growth is slow but ruin is rapid. Seneca was not the first to write about collapses, but he was probably the first who recognized that collapses are a fact of life. Collapses occur all the time, everywhere and, over your lifetime, you are likely to experience at least a few relatively large collapses: natural phenomena such as hurricanes, earthquakes, or floods. You may see large structures, buildings or bridges, crumble and major financial collapses such as the one that took place in 2008, as well as wars and social violence. And you may well see small-scale personal disasters such as losing your job or divorcing. You may not like that, but it is the way the universe works.

2. **Collapse is Rapid (the Seneca Cliff).** As Seneca noted, it takes only a short time for a large and apparently solid structure to unravel at the seams and crumble down in a heap. Think of the collapse of a house of cards, or that of the twin towers after the 9/11 attacks, or even of apparently slow collapses such as that of the Western Roman Empire: demise is always much faster than growth. Collapses are fast, it is one of their characteristics: you have to take that into account when you prepare for the worst.

3. **Collapse is Often Unexpected (the Seneca Peak).** Rarely does collapse give you an advance warning and some collapses are totally unpredictable, earthquakes, for instance. In other cases, the continuing growth before the crash may lull you to a false sensation of security, as it happened more than

© Springer Nature Switzerland AG 2020
U. Bardi, *Before the Collapse*,
https://doi.org/10.1007/978-3-030-29038-2

once to the fishing industry when the fish stocks collapsed just after that an all-time production high (the "Seneca Peak"). But, if you understand how and why collapses occur, you may at least be prepared for them and avoid their worst effects.

4. **Collapse is Bad for You (the Seneca Bottleneck)**. Collapses are a serious matter: they destroy things, kill people, generate sickness, make you sad, unhappy and depressed and, sometimes, they are irreversible. Yet, sometimes they are necessary to redress a situation that was impossible to control and they have to be accepted as a fact of life.

5. **There is Life After Collapse (the Seneca Rebound)**. Collapse is nothing but a "tipping point" from one condition to another. You can't go back but you can move onward and what looks like a disaster may be nothing but a passage to a new condition which may be better than the old one. This can be called the "Seneca Rebound," a characteristic of the evolution of complex systems. So, if you lose your job that may give you the opportunity to seek a better one. And if your company goes belly up, you may start another one without making the same mistakes you did with the first. Even disasters such as earthquakes or floods may be an opportunity to understand what is your role in life, as well as give you a chance to help your family and your neighbors.

6. **Resisting Collapse is Not a Good Idea (the Seneca Strategy)**. Collapse is the way the universe uses to get rid of the old to make space for the new. Resisting collapse means to strive to keep something old alive—you may succeed for a while, but often at the price of creating an even worse collapse. Often, you stick to your job, to your marriage, to your habits, as if your life were depending on not losing them, but you also know that, eventually, nothing can last forever. The Seneca Strategy consists in letting nature follow its course and let something go and disappear as it should. If you understand that, the bad effects of collapses can be reduced and, in some cases, you can even profit from them.

Lightning Source UK Ltd.
Milton Keynes UK
UKHW021836090420
361582UK00001B/2

9 783030 290375